双语动物学学习指导

Bilingual Zoology Study Guide

毛明光　蒋洁兰　主编

中国海洋大学出版社

·青岛·

图书在版编目（CIP）数据

双语动物学学习指导：中文、英文 / 毛明光，蒋洁兰主编. -- 青岛：中国海洋大学出版社，2022.6

ISBN 978-7-5670-3172-2

Ⅰ. ①双… Ⅱ. ①毛… ②蒋… Ⅲ. ①动物学—高等学校—教学参考资料—汉、英 Ⅳ. ①Q95

中国版本图书馆CIP数据核字(2022)第092781号

双语动物学学习指导

SHUANGYU DONGWUXUE XUEXI ZHIDAO

出版发行 中国海洋大学出版社

社　　址 青岛市香港东路23号　　**邮政编码** 266071

网　　址 http://pub.ouc.edu.cn

出 版 人 杨立敏

责任编辑 姜佳君

电子信箱 j.jiajun@outlook.com

电　　话 0532-85901040

印　　制 日照日报印务中心

版　　次 2022年6月第1版

印　　次 2022年6月第1次印刷

成品尺寸 210 mm×285 mm

印　　张 15.25

字　　数 362千

印　　数 1—1000

定　　价 59.80元

订购电话 0532-82032573（传真）

发现印装质量问题，请致电18663037500，由印刷厂负责调换。

《双语动物学学习指导》编写者名单

主　　编：毛明光（海南热带海洋学院）

　　　　　　蒋洁兰（海南热带海洋学院）

副 主 编：王海山（海南热带海洋学院）

参　　编：艾庆辉（中国海洋大学）

　　　　　　叶　乐（海南热带海洋学院）

　　　　　　霍忠明（大连海洋大学）

　　　　　　陈立婧（上海海洋大学）

　　　　　　王庆恒（广东海洋大学）

　　　　　　陈　治（海南热带海洋学院）

　　　　　　刘玉猛（大连海洋大学）

主　　审：茅云翔（海南热带海洋学院）

　　　　　　曹善茂（大连海洋大学）

前　　言

本书参照《普通动物学》（刘凌云等，2009）和*General Zoology*（Tracy I. Storer et al，1979），借鉴国外优秀开放资源编写而成。书中通过图解析动物学的基本原理，使读者更加容易理解和记忆，可作为生物科学类、畜牧业类、水产类，以及海洋生物学等相关专业双语动物学教与学的重要工具书。

对于高校教师，本书可作为教学参考资料，弥补传统教材的不足，为双语教学提供便利。对于高校相关专业的学生，本书可作为有利的学习工具。本书附有各章节重点词汇的注释与音标，并在章节后列出了学习要点，附上了试题，从而便于读者轻松掌握动物学的核心知识点。对于考研的同学，本书既可以增加专业词汇量，又可以提高外文阅读能力，书中附加的试题可以提高复习效率，达到事半功倍的效果。对于准备留学的同学，书中的图文可促进生词记忆，助力雅思、托福等考试的准备。

本书编写分工如下：第一章至第十章无脊椎动物部分以及第十八章动物进化部分由毛明光编写；第十一章至第十七章脊椎动物部分由蒋洁兰、毛明光、王海山、霍忠明、刘玉猛、叶乐、陈治编写；语法矫正、主要词汇的注释由蒋洁兰完成；巩固测验、试题库部分由毛明光、王海山编写。

曹善茂教授为本书的编写提供了大量珍贵外文资料，书中部分图片由高瑞、王树棋、涂宇涵 、王宇驰绘制，茅云翔教授为本书的编写与出版提供了指导和帮助，在此一并深表谢意！感谢海南热带海洋学院对本书出版的资助！

本书的编写是在新冠肺炎疫情期间完成的。由于我们的水平有限，书中的不妥之处在所难免，敬请读者批评指正。

毛明光

2021年10月

Contents

Chapter 1 Introduction

1–1 Definition of Zoology

Zoology, or animal biology, is the field of biology that involves the study of animals. The word "zoology" comes from the Greek words "zōion", meaning "animal", and "logos", meaning "study". It encompasses all aspects of scientific knowledge about animals, like embryonic development, evolution, behavior, ecological distribution, and classification. Zoology is broken into many branches because there are so many different ways to study animals; it is also broken into branches based on which animals are being studied.

1–2 History of Zoology

People have been interested in learning about animals since ancient times. The prominent ancient Greek philosopher Aristotle took detailed notes on animal observations, and inspired other scientists for many hundreds of years. Many universities were founded in Europe in the 16th century, and by the mid-17th century, divisions that focused entirely on animal research were founded in universities.

In the 19th century, the microscope became commonplace in scientific research, and this opened up a whole new realm of possibility; now, the cells of animals could be studied at the microscopic level. Another breakthrough in zoology occurred when the naturalist Charles Darwin developed the theory of evolution by natural selection. This theory revolutionized zoology and taxonomy (classification). More recently, the discovery of DNA as life's genetic material led to even more new research and knowledge about the natural world and the evolutionary relationships among animals.

1–3 Branches of Zoology

● **Zoography**

Zoography, also called descriptive zoology or zoogeography, is the study of animals and their habitats. It is concerned with the geographic ranges of specific populations of animals, their effects on the ecosystems they live in, and the reasons for a specific spatial distribution of an animal species.

- **Comparative Anatomy**

Comparative anatomy is the study of similarities and differences in the anatomy of different types of animals. Closely related animals like mammals share common bones, even if these bones have been extremely modified in shape. For example, the bones in bat wings are homologous to those in human fingers. Studying the similar anatomical structures in related organisms provided evidence for evolution from a common ancestor that was later confirmed by genetics research. Comparative anatomy is still used today, often in paleontology—the study of fossils.

- **Animal Physiology**

Animal physiology is the study of the bodily processes that occur in animals and allow them to maintain homeostasis and survive. Homeostasis is the ability of the body to maintain a relatively constant equilibrium even in a changing environment. One example is the regulation of body temperature in mammals. Humans have a normal body temperature of about 37°C (98.6°F), even when the environment they are in is much colder. Animal physiology involves the study of processes like temperature regulation, blood pressure and blood flow, and the release of hormones at specific times in the body.

- **Ethology**

Ethology is the study of animal behavior in their natural environment. Ethology has roots in the work of Darwin, but emerged as a field in the 1930s. It involves the study of animal learning, cognition, communication, and sexuality, and is related to evolutionary biology and ecology. Principles from ethology research are also used in animal training.

- **Behavioral Ecology**

Behavioral ecology emerged from ethology. It is the study of evolution as the basis for animal behavior due to ecological pressures, which are constraints placed upon organisms by their environment. Organisms with traits that are well-suited to their environment have a higher likelihood of surviving and reproducing than those who do not. When the proportion of individuals with favored traits increases over a long period of time, evolution can occur. Behavioral ecologists study animals' competition for resources such as food, territory, and mates and the increased reproductive success that certain traits may give.

1-4 Groupings by Animal

Zoology is also broken down into subcategories based on the type of animal being studied. For example, a distinction is made between invertebrate zoology and vertebrate zoology. There are also many specific terms for each type of animal that is studied. Some examples are:

Mammalogy: the study of mammals. A popular type of mammalogy is primatology—the study of primates.

Ornithology: the study of birds.

Herpetology: the study of amphibians and reptiles.

Ichthyology: the study of fish.

Entomology: the study of insects. Entomology is itself broken down into many categories because there are so many types of insects. Some examples of its subcategories are Lepidopterology (the study of butterflies and moths), Myrmecology (the study of ants), and Coleopterology (the study of beetles).

1–5 Definition of Taxonomy

Taxonomy is the branch of biology that classifies all living things. It was developed by the Swedish botanist Carl Linnaeus, who lived in the 18th century, and his system of classification is still used today. Linnaeus invented binomial nomenclature, the system of giving each type of organism a genus and species name. He also developed a classification system called the taxonomic hierarchy, which today has eight ranks from general to specific: domain, kingdom, phylum, class, order, family, genus, and species.

● **Taxonomic Hierarchy**

A taxon (plural: taxa) is a group of organisms that are classified as a unit. This can be specific or general. For example, we could say that all humans are a taxon at the species level since they are all the same species, but we could also say that humans along with all other primates are a taxon at the order level, since they all belong to the order Primates. Species and orders are both examples of taxonomic ranks, which are relative levels of grouping organisms in a taxonomic hierarchy. The following is a brief description of the taxonomic ranks that make up the taxonomic hierarchy.

(1) Domain

A domain is the highest (most general) rank of organisms. Linnaeus did invent some of the taxonomic ranks, but he did not invent the domain rank, which is relatively new. The term domain wasn't used until 1990, over 250 years after Linnaeus developed his classification system in 1735. The three domains of life are Bacteria, Archaea, and Eukarya. Archaea are single-celled organisms similar to bacteria. Some archaea live in extreme environments, but others live in mild ones. Eukaryota, or every living thing on earth that is not a bacterium or archaeon, is more closely related to the domain Archaea than to Bacteria.

(2) Kingdom

Before domains were introduced, kingdom was the highest taxonomic rank. In the past, the different kingdoms were Animalia, Plantae, Fungi, Protista, Archaea, and Bacteria (Archaea and

Bacteria were sometimes grouped into one kingdom—Monera). However, some of these groupings, such as Protista, are not very accurate. Protista includes all eukaryotic organisms that are not animals, plants, or fungi, but some of these organisms are not very closely related to one another. There is no set agreement on the kingdom classification, and some researchers have abandoned it altogether. Currently, it continues to be revised. In 2015, researchers suggested splitting Protista into two new kingdoms—Protozoa and Chromista. Currently there are five kingdoms in which all living things are divided: Monera, Protista, Fungi, Plantae, and Animalia (Fig. 1-1). Robert Whittaker (Fig. 1-2) was the first to propose the five-kingdom taxonomic classification in 1969.

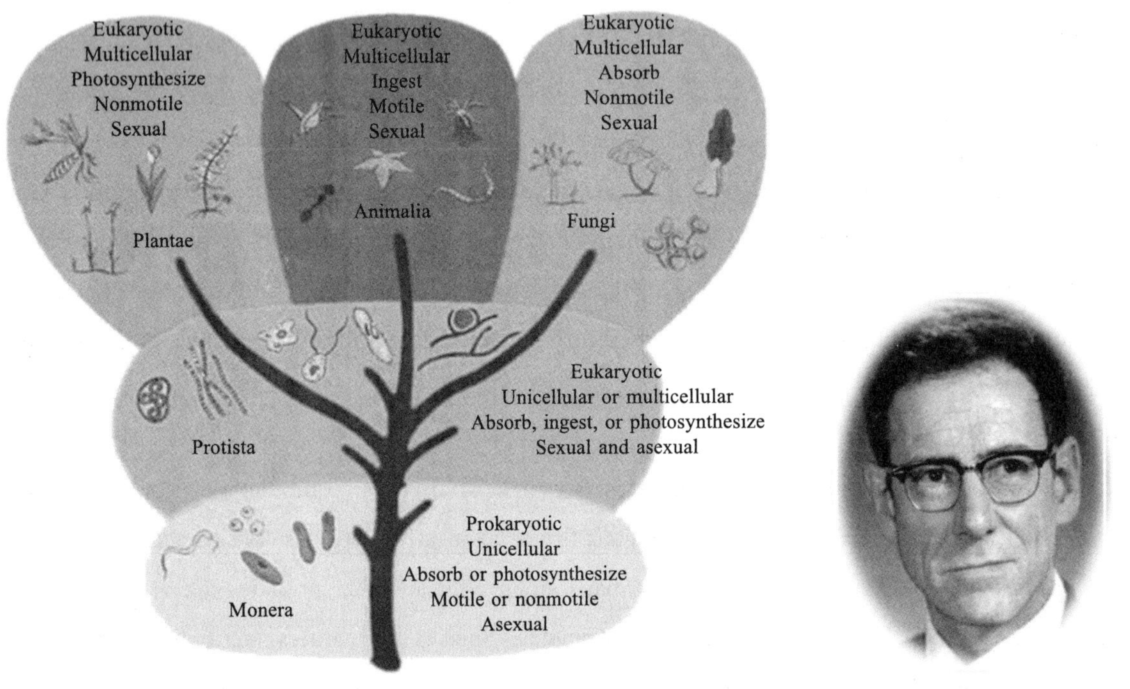

Fig. 1-1 Five-kingdom classification

Fig. 1-2 Robert Whittaker (1920–1980)

(3) Phylum

Phylum is the next rank after kingdom; it is more specific than kingdom, but less specific than class. There are about 40 phyla in the kingdom Animalia, including Chordata (all organisms with a dorsal nerve cord), Porifera (sponges), and Arthropoda (arthropods), and so on.

(4) Class

Class was the most general rank proposed by Linnaeus; phyla were not introduced until the 19th century. There are about 110 different classes in the kingdom Animalia, including Mammalia (mammals), Aves (birds), and Reptilia (reptiles), among many others. The classes of Animalia that Linnaeus proposed are similar to the ones used today, but Linnaeus' classes of plants were based on attributes like the arrangement of flowers rather than relatedness. Today's classes of plants are different from the ones Linnaeus used, and classes are not frequently used in botany.

(5) Order

Order is more specific than class. Some of Linnaeus' orders are still used today, such as Lepidoptera (the order of butterflies and moths). There are 19–26 orders of Mammalia, depending on how organisms are classified—sources differ. Some orders of Mammalia are Primates, Cetartiodactyla (whales, dolphins, porpoises, pigs, and giraffes), Carnivora (large carnivores/omnivores), and Chiroptera (bats).

(6) Family

Family is, in turn, more specific than order. Some families in the order Carnivora, for example, are Canidae (dogs, wolves, foxes), Felidae (cats), Mephitidae (skunks), and Ursidae (bears). There are about 15 total families in the order Carnivora.

(7) Genus

Genus is even more specific than family. It is the first part of an organism's scientific name using binomial nomenclature; the second part is the species name. An organism's scientific name is always italicized, and the genus name is capitalized while the species name is not. Genus and species are the only taxonomic ranks that are italicized. The scientific name for humans is *Homo sapiens*. *Homo* is the genus name, while *sapiens* is the species name. All other species in the genus *Homo* are extinct. Some were ancestral to humans, such as *Homo erectus*. Others lived at the same time, were closely related, and interbred with *Homo sapiens*, such as *Homo neanderthalensis*—the Neanderthals.

(8) Species

Species is the most specific major taxonomic rank. Species are sometimes divided into subspecies, but not all species have multiple forms that are different enough to be called subspecies. There are an estimated 8.7 million different species of eukaryotic organisms on Earth, but the vast majority have yet to be discovered and categorized. While each genus name is unique, the same species names can be used for different organisms. For example, *Ursus americanus* is the American black bear, while *Bufo americanus* is the American toad. The species name is always italicized, but never capitalized. It is the only taxonomic rank that is not capitalized. In scientific articles where the scientific name is used many times, it is abbreviated after the first full use by using just the first letter of the genus name along with the full species name. *Homo sapiens* is abbreviated to *H. sapiens.*

A species is a group of organisms that share a genetic heritage, are able to interbreed and create offspring that are also fertile. Different species are separated from each other by reproductive barriers. These barriers can be geographical, such as a mountain range separating two populations, or genetic barriers that do not allow for reproduction between the two populations. Scientists have changed their definition of a species several times throughout history. Species is one of the most specific classification that scientists use to describe animals. Scientists use binomial nomenclature to describe animals without the confusion of common names.

● **Binomial Nomenclature**

Binomial nomenclature is the system of scientifically naming organisms developed by Carl Linnaeus. Linnaeus published a large work—*Systema Naturae* (*The System of Nature*), in which Linnaeus attempted to identify every known plant and animal. This work was published in various sections between 1735 and 1758, and established the conventions of binomial nomenclature, which are still used today. Binomial nomenclature was established as a way to bring clarity to discussions of organisms, evolution, and ecology in general. Without a formalized system for naming organisms, the discussion of them, even between peers that speak the same language, becomes nearly impossible. The number of different colloquial names for a single species can be staggering.

Each scientific name in binomial nomenclature consists of two names, also called descriptors or epithets. The first word is the generic epithet and describes the genus that an animal belongs to. The second word is the specific epithet and refers to the species of the organism. Typically, the words have a Latin base and describe the genus or species with references to traits that are specific to the group. When written, the text of a scientific name is usually italicized, to clarify that it is a scientific name written in binomial nomenclature. The generic epithet is always capitalized, while the specific epithet is written in lower-case. In some older documents, both may be capitalized. Typically, the full name should be written out. However, when discussing many species of the same genus, the generic name is sometimes abbreviated to the first letter, still capitalized.

Thus, some animals like the giant panda, *Ailuropoda melanoleuca*, are in the genus *Ailuropoda* and their species name is *melanoleuca* (Fig. 1-3). Note the capitalization difference to distinguish between genus and species.

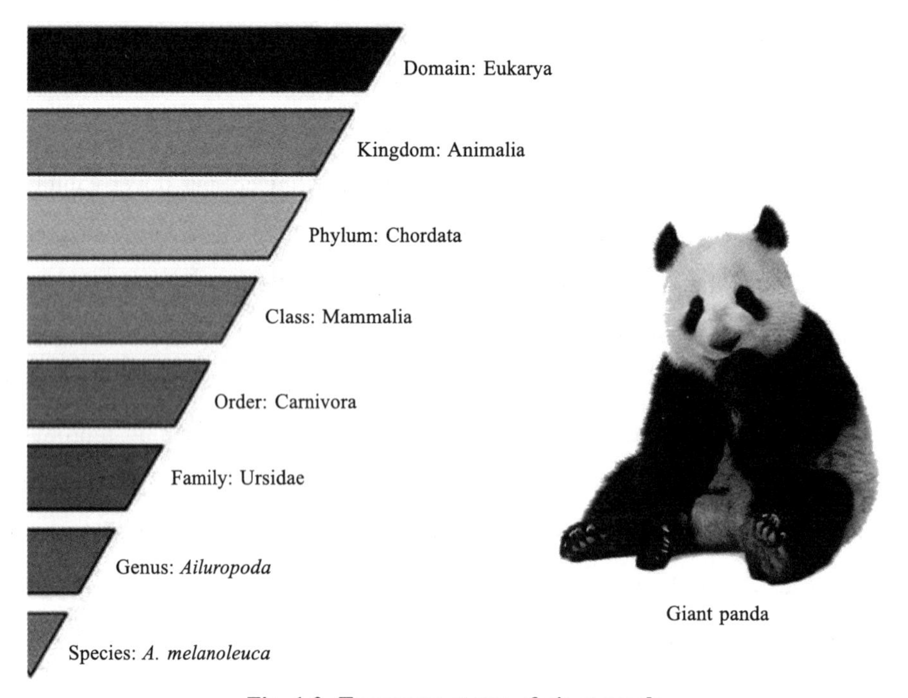

Fig. 1-3 Taxonomy status of giant panda

1-6 学习要点

● 动物学的定义：研究动物的形态、分类、生理、生态、分布、发生、遗传、进化及其与人类关系的科学。

● 延伸学科：动物形态学、动物分类学、动物生理学、动物胚胎学、动物遗传学、动物生态学、动物地理学等。

● 按研究类群分类：无脊椎动物学、脊椎动物学、原生动物学、蠕虫学、寄生虫学、贝类学、甲壳动物学、昆虫学、鱼类学、鸟类学、哺乳动物学（或兽类学）等。

● 研究动物学的目的、任务和方法。

● 动物资源的保护、动物资源的合理利用、有害动物的控制、实验动物。

● 动物学发展史（了解）。

● 动物界的分类及命名。

分类阶元是生物分类学确定共性范围的等级。现代生物分类采用的有域domain、界kingdom、门phylum、纲class、目order、科family、属genus、种species等8个阶元。

生物分界：五界系统、六界系统。

物种是具有一定形态特征和生理特性以及一定自然分布区的生物类群。一个物种中的个体一般不能与其他物种中的个体交配，或交配后不能产生有生殖能力的后代。

● 种的命名——双名法：属名+种名，斜体，属名首字母大写。

1-7 巩固测验

【名词】

物种、拉丁名、双名法、五界系统、阶元

【判断】

1735年，林奈以生物能否运动为标准，明确提出动物界和植物界的两界系统。

【选择】

国际动物命名法委员会规定：动物的学名为拉丁名，学名采用双名法，该名称是动物通用名。下列动物学名书写规范的是（　　）。

A. parus major　　B. *Parus Major*　　C. Parus major　　D. *Parus major*

【填空】

五界系统是指（　　　　）、（　　　　）、（　　　　）、（　　　　）、（　　　　）。

【简答】

1. 何为物种？为什么说它是客观的？

2. 双名法命名有何好处？它是怎样给物种命名的？

Chapter 2 Phylum Protozoa

2-1 Characteristics of Phylum Protozoa

- **Habitat:** mostly aquatic, either free living or parasitic or commensal.
- **Grade of organization:** protoplasmic grade of organization. Single cell performs all the vital activities, thus the single cell acts like a whole body. Body of protozoa is either naked or covered by a pellicle.
- **Locomotion:** Locomotory organ are pseudopodia (false foot) or cilia or absent.
- **Nutrition:** Nutrition are holophytic (like plant) or holozoic (like animal) or saprophytic or parasitic.
- **Digestion:** digestion is intracellular, occurs in food vacuoles.
- **Respiration:** through the body surface.
- **Osmoregulation:** Contractile vacuoles helps in osmoregulation.
- **Reproduction:** Asexually reproduction is through binary fission or budding. Sexual reproduction is by syngamy conjugation.
- **Classification of Protozoa:** Phylum Protozoa is classified into four classes on the basis of locomotary organs. They are Mastigophora, Sarcodina, Sporozoa, and Ciliata.

2-2 Class 1: Mastigophora/Flagellata

Flagellates are usually free living but few are parasitic forms. One or more flagella usually present for locomotion or food capturing or attachment or protection. Body is covered with a pellicle which provides a definite shape. Some forms are green due to the presence of chloroplasts (e. g. *Euglena*). Asexual reproduction occurs by longitudinal binary fission. Single nucleus presents in a cell.

- **Example:** ***Euglena***

All *Euglena* have chloroplasts and can make their own food by photosynthesis. They are not completely autotrophic though; *Euglena* can also absorb food from environment; *Euglena* usually live in quiet ponds or puddles. *Euglena* move by a flagellum (plural, flagella), which is a long whip-like structure that acts like a little motor (Fig. 2-1). The flagellum is located on the anterior (front) end, and twirls in such a way as to pull the cell through the water. It is attached at an inward pocket called the reservoir.

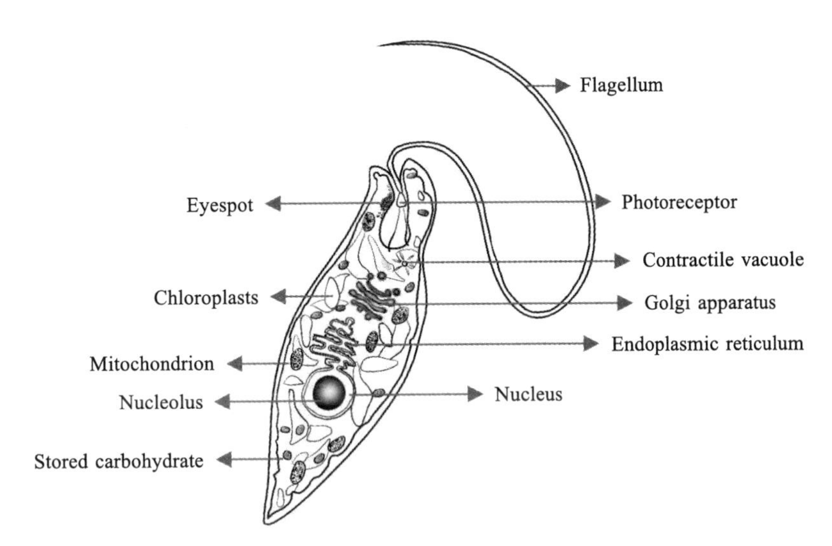

Fig. 2-1 Schematic of *Euglena*

The *Euglena* is unique in that it is both heterotrophic (must consume food) and autotrophic (can make its own food). Chloroplasts within the *Euglena* trap sunlight that is used for photosynthesis, and can be seen as several rod-like structures throughout the cell. *Euglena* also have an eyespot that detects light at the anterior end, which can be seen near the reservoir. This helps the *Euglena* find bright areas to gather sunlight to make its food. *Euglena* can also gain nutrients by absorbing them across their cell membrane, hence they become heterotrophic when light is not available, and they cannot photosynthesize. The *Euglena* has a stiff pellicle outside the cell membrane that helps it keep its shape, though the pellicle is somewhat flexible and some *Euglena* can be observed scrunching up and moving in an inchworm type fashion.

In the center of the cell is the nucleus, which contains the cell's DNA and controls the cell's activities. The nucleolus can be seen within the nucleus. The interior of the cell contains a jelly-like fluid substance called cytoplasm. Near the posterior of the cell is a star-like structure: the contractile vacuole. This organelle helps the cell remove excess water, and without it the *Euglena* could take in too much water due to osmosis that the cell would explode.

Some species, especially *E. viridis* and *E. sanguinea*, can develop large toxic populations of green or red "blooms" in ponds or lakes with high nitrogen content. *E. gracilis* is common in laboratory demonstrations, and a number of species are used to study cell growth and metabolism in various environmental conditions.

2-3 Class 2: Sarcodina

Sarcodina is characterized by the formation of pseudopods for locomotion and taking food.

● **Example:** ***Amoeba proteus***

The name "ameba" comes from the Greek word "amoibe", which means "change". Protists are microscopic unicellular organisms that do not fit into the other kingdoms. Some protozoans are considered plant-like, such as algae, and others are considered animal-like. The ameba is considered an animal-like protist because it moves and consumes its food, but it is not classified as an animal because it consists of a single cell; it is unicellular.

Protists are also classified by how they move. Some have cilia or flagella, but the ameba has an unusual way of creeping along by stretching its cytoplasm into fingerlike extensions called pseudopodium (Fig. 2-2). The word "pseudopodium" means "false foot". The cell membrane is very flexible and allows for the ameba to change shape. There are two types of cytoplasm in the ameba: the darker cytoplasm toward the interior of the protozoan is called endoplasm, and the clearer cytoplasm that is found near the cell membrane is called ectoplasm. By pushing the endoplasm toward the cell membrane, the ameba causes its body to extend and creep along. It is also by this method that the ameba consumes its food. The pseudopodia extend out and wrap around a food particle in a process called phagocytosis. The engulfed food then becomes a food vacuole. The food will eventually be digested by the cell's lysosomes.

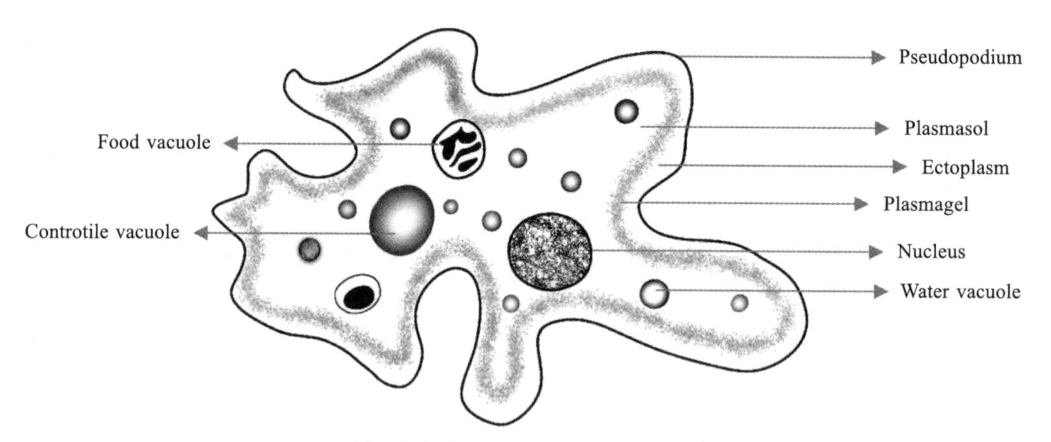

Fig. 2-2 Schematic of *Amoeba proteus*

Also visible in the ameba is the nucleus, which contains the ameba's DNA. In order to reproduce, the ameba goes through mitosis, when the nucleus duplicates its genetic material and the cytoplasm splits into two new daughter cells, each identical to the original parent. This method of reproduction is called binary fission. Another structure easily seen in the ameba is the contractile vacuole, whose job is to pump out excess water so that the ameba does not burst. Under unfavorable conditions, the ameba can create a cyst. This hard-walled body can exist for a long period of time until conditions become favorable again. At this point it opens up and the ameba emerges. Often cysts are created during cold or dry periods where the ameba could not survive in its normal condition.

Amebas can cause disease. A common disease caused by the ameba is called amebic dysentery. A person becomes infected by drinking contaminated water. The ameba then upsets the person's digestive system and causes cramps and diarrhea. A person is most likely to be infected in countries where the water is not filtered or purified.

2-4 Class 3: Sporozoa

Parasitic protozoans are usually immobile. They includes plasmodia, coccidia, piroplasm, and malaria parasites.

● **Characteristic**

All sporozoans are endoparasites. Some sporozoans such as *Eimeria* cause severe diseases like coccidiosis in the birds. Locomotory organelles (cilia, flagella, pseudopodia, etc.) are absent. Nutrition is parasitic (absorptive). Phagotrophy is rare. The body is covered with an elastic pellicle or cuticle. Contractile vacuoles are absent. Asexual reproduction occurs through multiple fission. Sexual reproduction takes place through syngamy. Life cycle consists of two distinct asexual and sexual phases. They may be passed in one (monogenetic) or two different hosts (digenetic).

● **Example: *Plasmodium vivax***

Plasmodium is an intracellular sporozoan parasite causing malaria in man. The parasite lives in the red blood cells and liver cells of man and alimentary canal and salivary glands of female *Anopheles* mosquitoes.

● **Structure**

Structure of *Plasmodium* is different among stages of its life cycle. A fully grown malarial parasite is an amoeboid and uninucleated structure known as trophozoite. Trophozoite is surrounded by double layered plasma lemma. Cytoplasm contains Palade granules, endoplasmic reticulum, ribosome, mitochondria, vesicles and vacuoles having haemozoin. Cytoplasm contains nucleus having nucleolus and granular nucleoplasm.

● **Life cycle**

The life cycle of *Plasmodium* can be divided into three phases: asexual schizogony, sexual gamogony, asexual sporogony (Fig. 2-3).

(1) Asexual Cycle of *Plasmodium* in Man

Asexual schizogony: Schizogony is an asexual phase of reproduction of *Plasmodium*. It takes place in liver cells and red blood cells of man. Schizogony can be divided into the following phases: pre-erythrocytic schizogony, exo-erythrocytic schizogony, erythrocytic schizogony, post-erythrocytic schizogony.

Sometimes, some merozoites produced in erythrocytic schizogony reach the liver cells and undergo schizogony development in liver cells. This is called post-erythrocytic schizogony.

(2) Sexual Cycle of *Plasmodium* in Mosquito

When a female *Anopheles* sucks the blood of a malaria patient, the gametocytes reach the stomach of the mosquito and formation of gametes takes place as follows. (a) Gametogenesis (gametogony). (b) Fertilization: The male gamete enters the female gamete through the fertilization cone formed at the female gamete and form diploid zygote or synkaryon. Fusion is anisogamous. (c) Ookinete stage: The zygote remains inactive for some time and then elongates into a worm-like ookinete or vermicule, which is motile. The ookinete penetrates the stomach wall and comes to lie below its outer epithelial layer. (d) Oocyst stage: The ookinete gets enclosed in a cyst. The encysted zygote is called oocyst. The oocyst absorbs nourishment and grows in size.

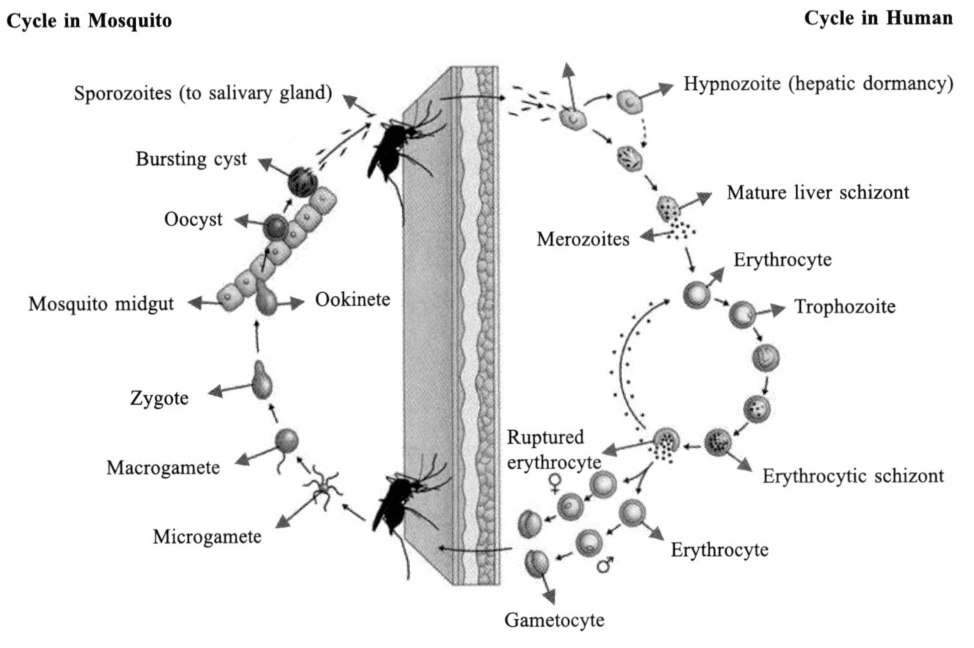

Fig. 2-3 Life cycle of *Plasmodium*

(3) Asexual Sporogony

The nucleus of oocyst divides repeatedly to form a large number of haploid daughter nuclei. At the same time, the cytoplasm develops vacuoles and gives numerous cytoplasmic masses. The daughter nuclei pass into each cytoplasmic mass and develop into slender sickle-shaped sporozoites in each oocyst. This phase of asexual multiplication is known as sporogony.

Lastly, the oocyst bursts and sporozoites are liberated into the hemolymph of the mosquito. They spread throughout the hemolymph and eventually reach the salivary glands and enter the duct of the hypopharynx. The mosquito is now becoming infective and sporozoites get inoculated or injected into the human blood when the mosquito bites. The cycle is repeated.In mosquito, the whole sexual cycle is completed in 10–12 days.

2-5 Class 4: Ciliata

Class of protozoa having cilia or hair-like appendages on part or all of the surface during some part of the life cycle.

● **Example: *Paramecium caudatum***

Paramecium is a unicellular protozoan classified in the phylum Ciliophora and the kingdom Protista. They live in quiet or stagnant ponds and are an essential part of the food chain. They feed on algae and other microorganisms, and other small organisms eat them. All members of the phylum Ciliophora move by tiny hair-like projections called cilia (Fig. 2-4, Fig. 2-5). *Paramecium* cannot change its shape like the ameba because it has a thick outer membrane called the pellicle. The pellicle surrounds the cell membrane.

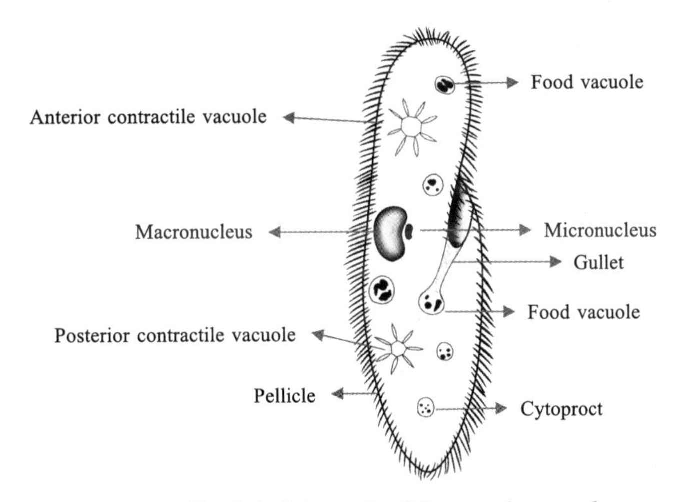

Fig. 2-4 Schematic of *Paramecium caudatum*

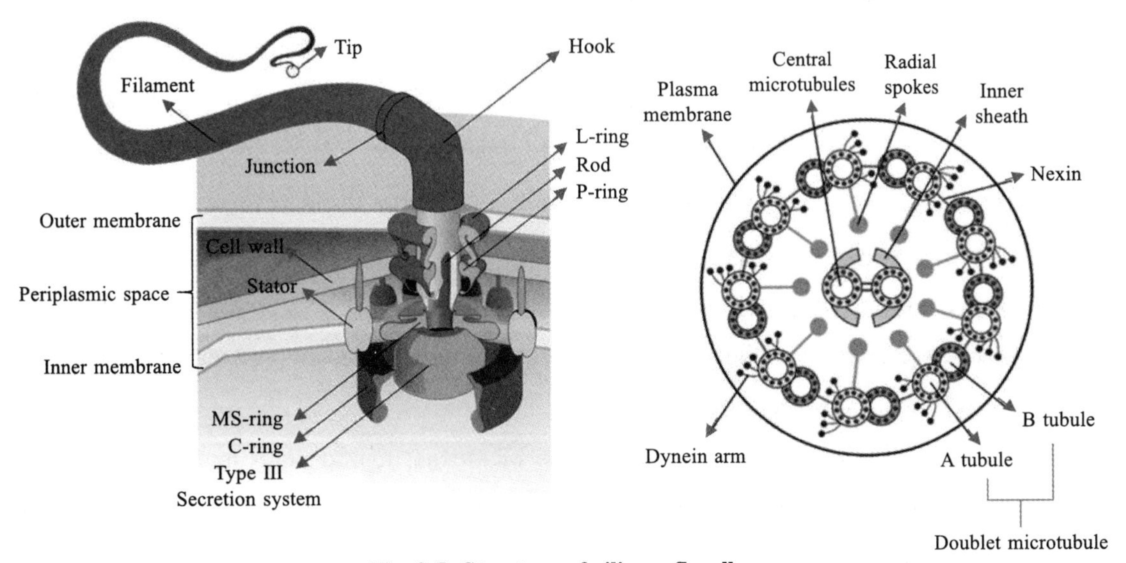

Fig. 2-5 Structure of cilia or flagellum

There are two types of nuclei. The large nucleus is called the macronucleus which controls cell activities such as respiration, protein synthesis and digestion. The much smaller micronucleus is used only during reproduction. Reproduction in *Paramecium* involves the exchanging of DNA within the micronucleus. In order to do this, two paramecia lie side by side and join at the mouth pore. This process is called conjugation and is a method of sexual reproduction in other microorganisms. Contractile vacuoles are used in animal cells to remove the excess water. The contractile vacuole is shaped like a star. Paramecia are heterotrophs, meaning they must consume food for their energy. Food enters *Paramecium* through the mouth pore and goes to the gullet. The area of *Paramecium* appears pinched inward and is called the oral groove. Cilia sweep food into this area. At the end of the gullet, food vacuoles are formed. Food vacuoles then remain in the cytoplasm until the food is digested. Undigested food particles are eliminated through the anal pore.

Paramecium can respond to temperature, food, oxygen and toxins and has a very simple defense mechanism. Just inside the pellicle are threadlike organelles called trichocysts. The *Paramecium* can shoot tiny threads out of the cell to entangle a predator or to make themselves appear bigger. *Paramecium* is also known to exhibit avoidance behavior—they will move away from a negative or unpleasant stimulus.

There are two kinds of cytoplasm in the *Paramecium*. The cytoplasm around the edges is clear and is called ectoplasm. The rest of the cytoplasm is denser and appears darker. This is called the endoplasm. Remember that the word "ecto" means outside, and the word "endo" means inside.

● **Nutrition of *P. caudatum***

In *Paramecium caudatum*, nutrition is holozoic. The food comprises chiefly bacteria and minute protozoa. *Paramecium* does not wait for the food but hunts for it actively.

● **Respiration and Excretion of *P. caudatum***

The exchange of gases (oxygen and carbon dioxide) takes place through the semi-permeable pellicle like other freshwater protozoans by the process of diffusion. *P. caudatum* obtains its oxygen from the surrounding water. Carbon dioxide and organic wastes like ammonia resulting from metabolism are probably excreted by diffusing outward into the water in the reverse direction.

● **Reproduction in *P. caudatam***

P. caudatum reproduces asexually by transverse binary fission and also undergoes several types of nuclear re-organization, such as conjugation, endomixis, autogamy, cytogamy and hemixis, etc.

(1) Transverse Binary Fission

Transverse binary fission is the commonest type of asexual reproduction in *Paramecium*. It is a distinctly unique asexual process in which one fully grown specimen divides into two daughter individuals without leaving a parental corpse (Fig. 2-6).

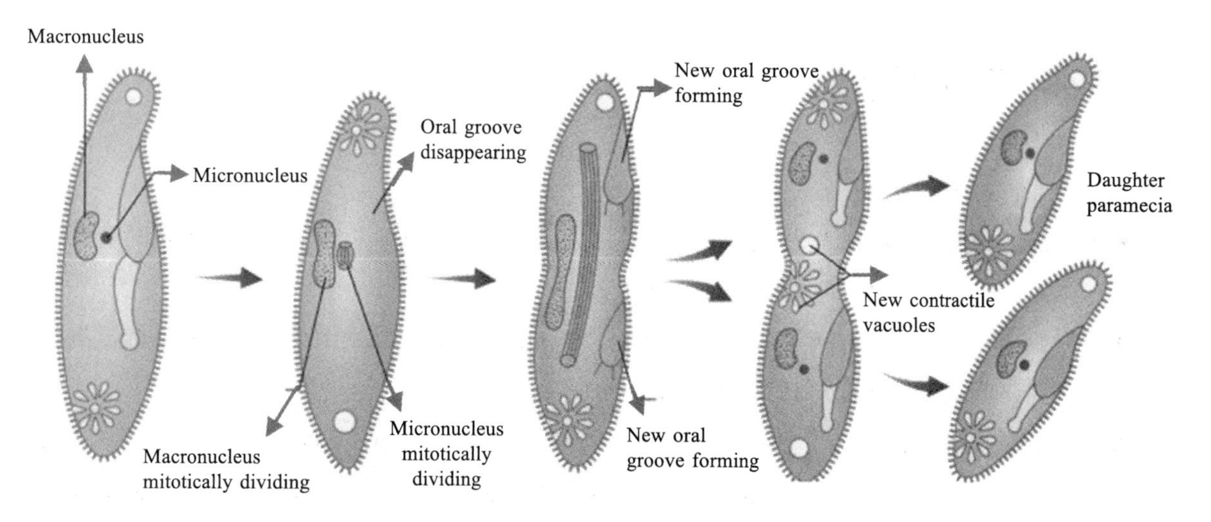

Fig. 2-6 Transverse binary fission

(2) Conjugation

Ordinarily *P. caudatum* multiplies by binary fission for long periods of time, but at intervals this may be interrupted by the joining of two animals along their oral surfaces for the sexual process of conjugation. Conjugation is defined as the temporary union of two individuals which mutually exchange micronuclear material. It is a unique type of a sexual process in which two organisms separate soon after the exchange of nuclear material (Fig. 2-7).

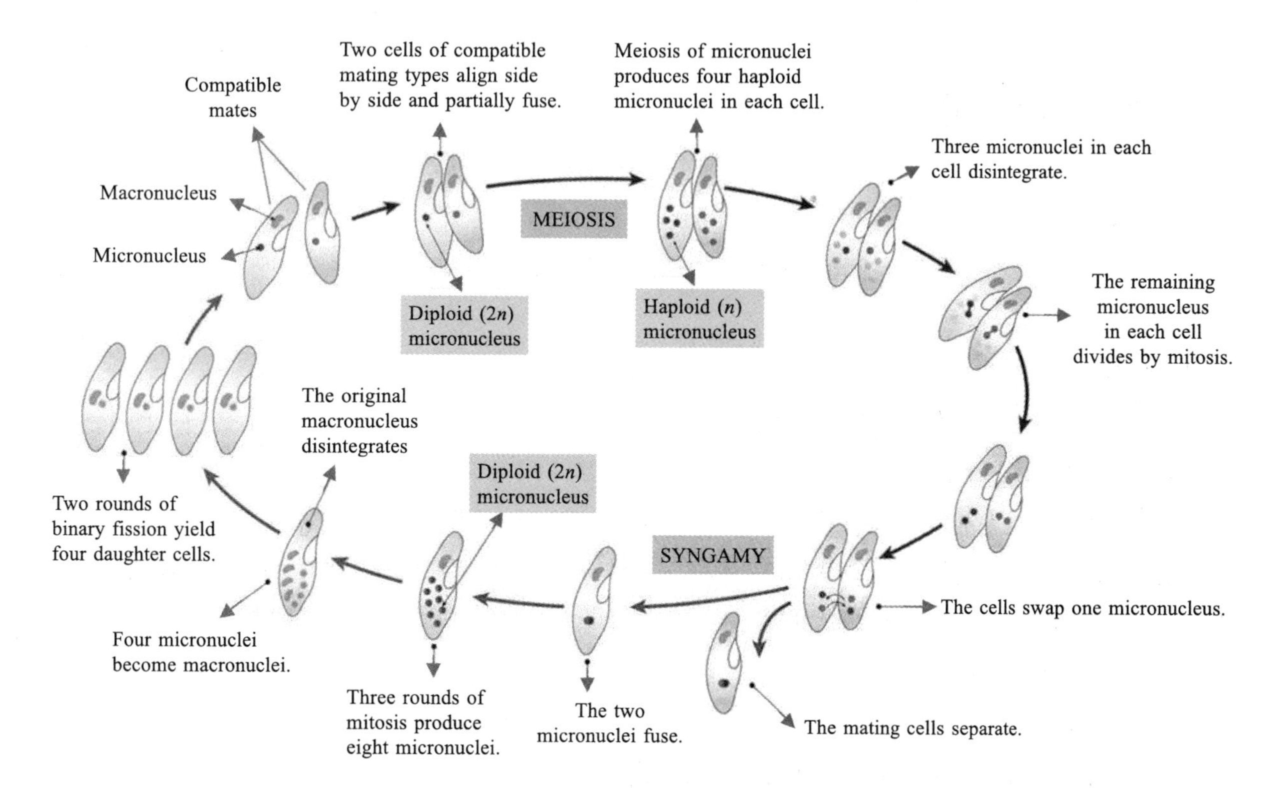

Fig. 2-7 Conjugation

2-6 学习要点

● 原生动物门的主要特征。

● 鞭毛纲 Mastigophora 的代表动物——眼虫 *Euglena*；鞭毛纲的主要特征；鞭毛纲的重要类群（植鞭亚纲 Phytomastigina、动鞭亚纲 Zoomastigina）。

● 肉足纲 Sarcodina 的代表动物——大变形虫 *Amoeba proteus*；肉足纲的主要特征；肉足纲的重要类群（根足亚纲 Rhizopoda、辐足亚纲 Actinopoda）。

● 孢子纲 Sporozoa 的代表动物——间日疟原虫 *Plasmodium vivax*；间日疟原虫在人体内进行裂体生殖（红细胞前期 pre-erythrocytic stage、红细胞外期 exo-erythrocytic stage、红血细胞内期 erythrocytic stage），在按蚊体内进行配子生殖和孢子生殖；孢子纲的主要特征；孢子纲的重要类群。

● 纤毛纲 Ciliata 的代表动物——大草履虫 *Paramecium caudatum*；纤毛纲的主要特征；纤毛纲的常见种类。

● 鞭毛/纤毛的超显微结构；伪足形成的机制。

● 草履虫的接合生殖。

● 疟原虫的生活史及其世代交替现象和合子减数分裂。

● 草履虫、眼虫的结构简图。

2-7 巩固测验

【名词】

胞吞、胞吐、胞饮、孤雌生殖、包囊、胞遗、滋养体、裂殖体、接合生殖、刺丝泡、伪足、世代交替

【选择】

1. 原生动物的伸缩泡最主要的作用是（　　）。

A. 调节水分平衡　　B. 排除代谢废物

C. 排出未消化的食物残渣　　D. 进行气体交换

2. 眼虫制造过多食物常形成一些半透明的（　　）贮存于细胞质中，这是眼虫的特征之一。

A. 淀粉粒　　B. 副淀粉粒　　C. 食物泡　　D. 叶绿体

3. 能形成赤潮的原生动物是（　　）。

A. 草履虫　　B. 绿眼虫　　C. 痢疾内变形虫　　D. 腰鞭毛虫

4. 草履虫是研究动物生物学的好材料，其体内大核的功能是（　　）。

A. 孤雌生殖　　B. 营养与代谢　　C. 纵二分裂　　D. 横二分裂

5. 能行光合作用的原生动物是（　　）。

A. 变形虫　　B. 草履虫　　C. 绿眼虫　　D. 疟原虫

6. 变形虫最常见的生殖方式是（　　）。

A. 孢子生殖　　B. 配子生殖　　C. 二分裂生殖　　D. 出芽生殖

7. 刺丝泡为下列动物中（　　）所特有。

A. 眼虫　　B. 草履虫　　C. 变形虫　　D. 孢子虫

8. 间日疟原虫的中间寄主为（　　）。

A. 钉螺　　B. 沼螺　　C. 按蚊　　D. 伊蚊

9. 疟原虫在蚊体内进行的生殖方式是（　　）。

A. 配子生殖和孢子生殖　　B. 孢子生殖和裂体生殖

C. 裂体生殖和配子生殖　　D. 配子生殖和分裂生殖

10. 接合生殖属于（　　）。

A. 孤雌生殖　　B. 有性生殖　　C. 无性生殖　　D. 孢子生殖

【简答】

1. 为什么说原生动物是最原始、最低等的一类动物?

2. 原生动物门的主要特征是什么？它有哪几个纲？分纲的主要依据是什么？

3. 能引起水华现象的原生动物有哪些?

Chapter 3 Phylum Porifera/Spongia

3-1 Characteristics of Phylum Porifera/Spongia

- Their bodies consist of loosely organized cells.
- They vary in size from one centimeter to one meter.
- They are asymmetrical or radially symmetrical.
- They have three cell types: pinacocytes, mesenchymal cells, and choanocytes.
- They have central cavity or spongocoel. This cavity may be divided into series of branching chambers. Water circulates through these chambers for feeding.
- Numerous pores are present in the body wall, i.e. ostia and oscula.
- They have no tissue or organ. Skeleton is composed of spicules.
- Nervous system is absent but neurosensory cells are present.
- Asexual reproduction takes place by budding.
- They are hermaphrodites and larvae are produced during development.

The word "porifera" mainly refers to the pore bearers or pore-bearing species. Based on the embryological studies, sponges are proved as animals and are classified into a separate phylum in the animals.

This phylum includes about 5,000 species. Poriferans are the earliest multicellular animals. The pores are known as ostia. The poriferans have a spongy appearance and are therefore called sponges. They are attached to the substratum and do not move. They have the ability to absorb and withhold fluids. They were initially regarded as plants due to the green colour and their symbiotic relationship with algae. Later, their life cycle and feeding system were discovered and they were included in the animal kingdom.

- **Cell Types, Body Wall and Skeletons**

Sponges have simple bodies. But still sponges are more than colonies of independent cells.

Sponges also have specialized cells. Therefore, division of labour is present in them. The following types of cells are present in phylum Porifera.

(1) Pinacocytes

Pinacocytes are thin walled and flat cells (Fig. 3-1). They line the outer surface of a sponge. Pinacocytes are slightly contractile. Their contraction can change the shape of some sponges. Some pinacocytes forms tube-like contractile porocytes. Porocytes regulate water circulation. The openings of the porocytes are pathways of water through the body wall.

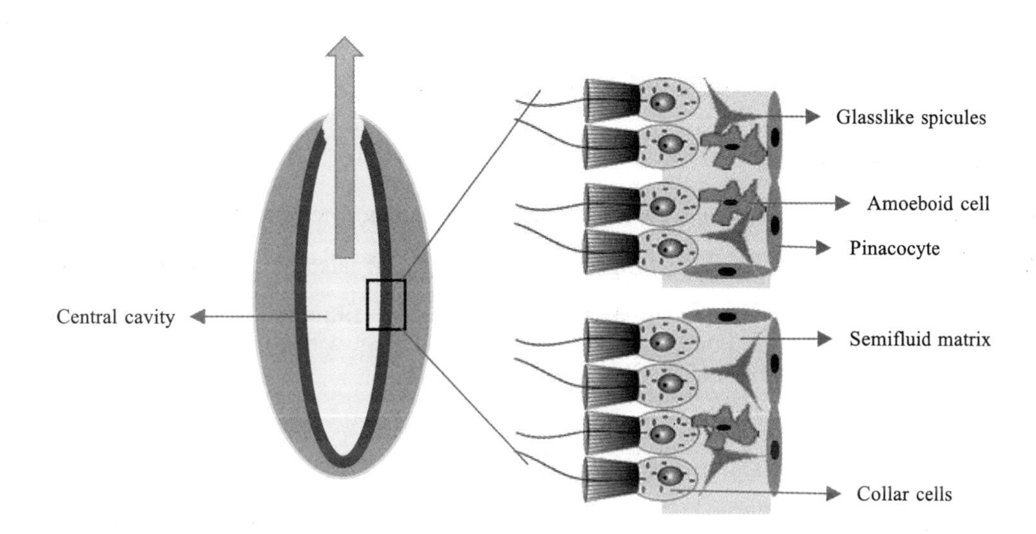

Fig. 3-1 Cell types of sponges

(2) Mesohyl

Mesohyl is a jelly-like layer present below the pinacocytes. Amoeboid cells are present in it. These cells are called mesenchyma cells. The mesenchymal cells freely move in the mesohyl. These cells are specialized for reproduction, secreting, skeletal elements, transporting and storing food and forming contractile rings around openings in the sponge wall.

(3) Choanocytes

Choanocytes or collar cells are present below the mesohyl. They form the lining of the inner chamber. Choanocytes are flagellated cells. They have a collar-like ring of microvilli surrounding a flagellum. Microfilaments connect the microvilli. It forms a netlike structure within the collar. The flagellum creates water currents through the sponge. The collar filters microscopic particles from the water. Collar cells are also present in a group of protists called choanoflagellates. It suggests an evolutionary link between these groups.

(4) Skeleton

The nature of the skeleton is an important characteristic in sponge taxonomy. There are two types of skeleton in sponges.

The spicules consist of microscopic needlelike spikes, which are formed by amoeboid cells. They are made of calcium carbonate or silica. They have different shapes (Fig. 3-2).

The spongin fibers are made up of spongin. Spongin is a fibrous protein made of collagen.

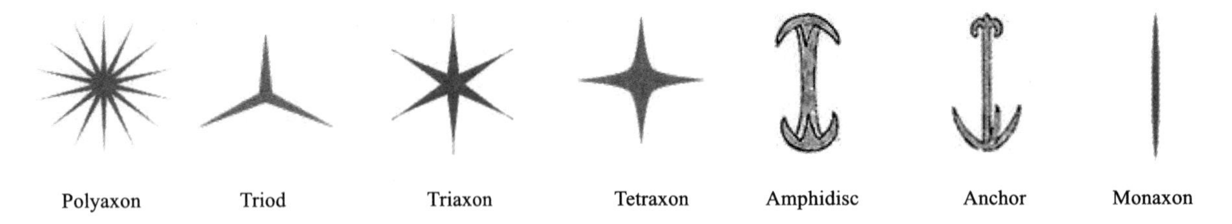

Fig. 3-2 Different spicule shapes of sponges

● **Water Canal System**

The life of a sponge depends on the water currents. The choanocytes organize to form water canal system. Water currents bring food and oxygen for a sponge. It also carries away metabolic and digestive wastes. Water canal system is used for circulation and filtration of food. There are three principal types of water canal systems (Fig. 3-3).

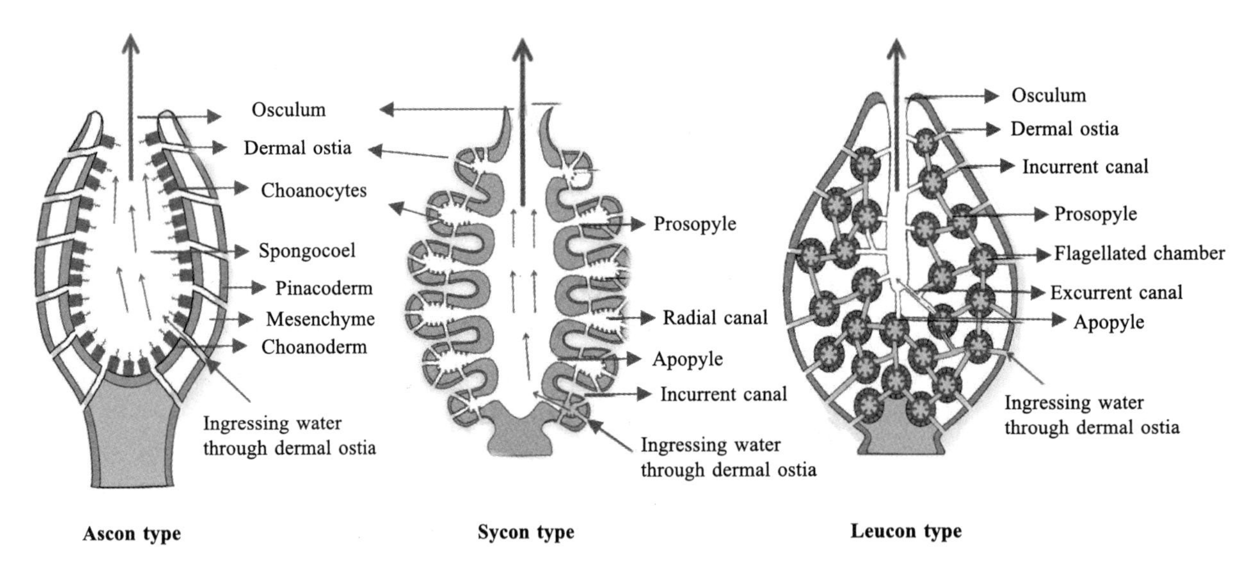

Fig. 3-3 Water canal systems of sponges

(1) Ascon Type

It is the simplest and least common sponge body form. Ascon sponges are vase-like. Ostia are the outer openings of porocytes. The porocytes open directly into spongocoel. Choanocytes line the spongocoel. The movements of flagella of choanocytes draw water into the spongocoel through the ostia. Water leaves the sponge through the osculum. Osculum is a single large opening at the top of the sponge.

(2) Sycon Type

The sponge wall is folded in the sycon body form. The following canals are formed by the folding of its wall. (a) Incurrent canal: The invaginations of the body wall of sycon form incurrent canals. Water enters into incurrent canal through dermal pores. (b) Radial canal: Pores in the wall of incurrent wall connect incurrent canals with radial canals. Choanocytes line the radial canals. The beating of flagella of choanocytes moves water through incurrent radial canals and spongocoel. It finally moves out the osculum. (c) Spongocoel: The radial canals lead to spongocoel. Water path through sycon type: dermal pore–incurrent canal–pore–radial canal–spongocoel–osculum.

(3) Leucon Type

Leucon sponges have an extensively branched canal system. There are the following chambers in leucon type. (a) Branched incurrent canal: Water enters the branched incurrent canals through ostia. (b) Choanocytes chamber: Incurrent canals lead into choanocytes-lined chambers. (c) Excurrent canal: Choanocyte chamber open into the chambers of excurrent canals. A large number of chambers and canals are present in leucon type. Therefore, spongocoel is absent in them. They have many oscula for water leaving the sponge. Leucon type canal system is formed by the evolution of simple canal system.

Advantages of complex canal system: Complex sponges have an increased surface area for choanocytes. Therefore, a large amount of water moves through the sponge. It increases the filtering capability of sponges.

● **Reproduction**

Reproduction for sponges can be accomplished both sexually and asexually. There are three ways for a sponge to reproduce asexually: budding, gemmules, and regeneration. Sponges can simply reproduce by budding, where a new sponge grows from the older one and eventually breaks off. Sponges can also reproduce by regeneration, where missing body parts are regrown. People who harvest sponges often take advantage of this by breaking off pieces of their catch and throwing them back in the water to be harvested later. Finally, sponges can reproduce by creating gemmules which are a group of amebocytes covered by a hard outer covering (Fig. 3-4).

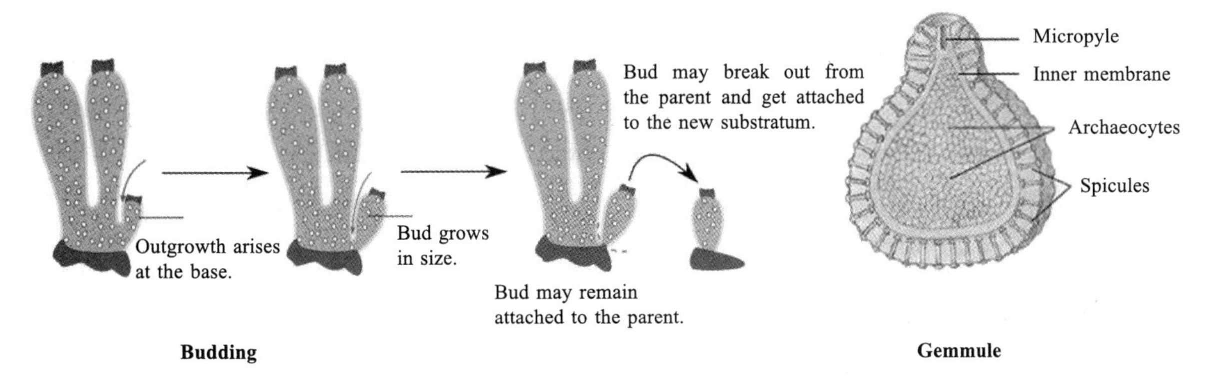

Fig. 3-4 Reproduction of sponges

Sexual reproduction occurs when one sponge releases sperm into the water. This sperm travels to another sponge and fertilizes its egg. The larva form will then swim to another location using its flagella where it will grow into an adult sponge. Most sponge species are hermaphrodites—they can produce both eggs and sperms. Sperms are released out from sponges through the outgoing water from osculum. The sperms thus make their way into another sponge through incoming water by ostia. Choanocytes act as nurse cells and transport the sperms to the ova which lie in the flagellated choanoderm. The fertilization is internal and cross type. Early development takes place within the maternal sponge leading to the formation of larval stages. The larval stages bear flagella, which help them to escape out from the maternal sponge body. The larva thus gets attached to a suitable substratum, metamorphoses and grows into an adult sponge. Sponges have two types of larvae (Fig. 3-5).

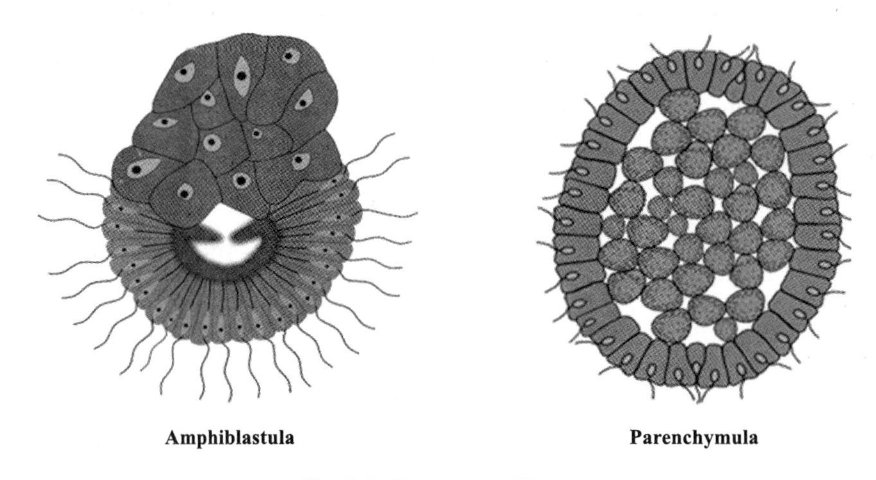

Fig. 3-5 Two types of larvae

(1) Amphiblastula

It is hollow, oval larval stage characteristic of calcareous sponges (Scypha). The anterior half of amphiblastula bears flagella while the posterior half is free from flagella.

(2) Parenchymula

It is solid, oval or flattened larval stage characteristic of calcareous sponges, Hexactinellida and most Desmospongia. The entire larva is covered by flagella.

With the help of external flagella, the motile larvae escape from the parental body and swim for a few hours to many days. Finally, they settle down, become attached to some solid objects, metamorphose and grow into an adult.

3-2 学习要点

● 海绵动物的形态结构［不对称体形 asymmetric body plan，非明确的组织、器官和系统 nonspecific tissues, organs and systems，水沟系 canal system（单沟型 ascon type、双沟型 sycon type 和复沟型 leucon type）]。

● 海绵动物的生殖和发育。

● 海绵动物门的分类及分类地位（钙质海绵纲 Calcarea、六放海绵纲 Hexactinellida、寻常海绵纲 Demospongiae）。

3–3 巩固测验

【名词】

领细胞、两囊幼虫、胚胎逆转、水沟系

【填空】

海绵动物水沟系的类型有（　　　　　　　　）、（　　　　　　　　）、（　　　　　　　　）。

【简答】

为什么说多孔动物是最原始、最低等的多细胞动物?

Chapter 4 Phylum Coelenterata/Cnidaria

4-1 Characteristics of Phylum Coelenterata/Cnidaria

- They are aquatic organisms, mostly are marine and few inhabit in freshwater.
- They are diploblastic animals with a radially symmetrical body.
- The body wall consists of outer ectoderm and inner endoderm or gastrodermis. Between the two layers, a non-cellular mesoglea is present.
- The body bears one central body cavity or gastrovascular cavity. It is known as the coelenteron which gives the animals the alternative name—Coelenterata.
- The representatives of cnidarians can make microscopic intracellular stinging capsules, known as nematocysts or cnidae, which also give the phylum name "Cnidaria".
- The ectodermal layer contains different types of cells like endothelio-muscle cells, interstitial cells, glandulo-muscle cells, cnidoblasts or nematocysts, sensory cells, germ cells, and nerve cells.
- Endoderm or gastrodermis contains five types of cells such as endothelio-muscle cells or nutritive cells, interstitial cells, endothelio-gland cells, nerve cells, and sensory cells.
- The interstitial cells have totipotent functions because they have the capability to produce any other cells like reproductive cells, glandular cells, and cnidoblasts which secrete hypnotoxin (poison).
- They have incomplete alimentary canal because the anal opening is absent whereas the gut cavity contains a single opening which is known as mouth. In this case, mouth serves for ingestion and egestion.
- They are carnivorous and can digest all types of food. In this case, digestion is unique with extracellular digestion followed by intracellular digestion.
- They have a short and slender body with tentacles which encircle the mouth in one or two whorls. In this case, tentacles function for food-engulfing, capturing the prey and defense.

- In nature, they have two distinct forms such as polyp and medusa (polymorphism). In this case, polyps are sessile (asexual stage) while medusa is free-swimming (sexual stage).
- In their body, circulatory, respiratory and excretory systems are absent. Through general body surface, gaseous exchange and excretion occur.
- They can move with the help of smooth muscle fibrils in the epithelia or ectoderm. Tentacles also play a vital role in the movement.
- The nervous system is poorly developed which makes the network of nerve fibers in the body walls and the tentacles.
- The body bears single or complicated sensory organs with eyespots called statocysts.
- They reproduce either asexual by external budding or sexual by the formation of ova and sperms.
- They show alternation of generation (metagenesis) in their life cycle, in which asexual colonial polypoid generation alternates with sexual free-swimming medusoid generation.
- Fertilization is internal or external and development is indirect with a larval stage. In this case, planula larva is common but ephyra larva is found in some animals.

Some cnidarians are polymorphic, having two body plans during their life cycle (Fig. 4-1). An example is the colonial hydroid called *Obelia*. The sessile polyp form has, in fact, two types of polyps. One is the gastrozooid, which is adapted for capturing prey and feeding; the other is the gonozooid, adapted for the asexual budding of medusa. When the reproductive buds mature, they break off and become free-swimming medusas, which are either male or female (dioecious). The male medusa makes sperms, whereas the female medusa makes eggs. After fertilization, the zygote develops into a blastula and then into a planula larva. The larva is free swimming for a while, but eventually attaches and a new colonial reproductive polyp is formed.

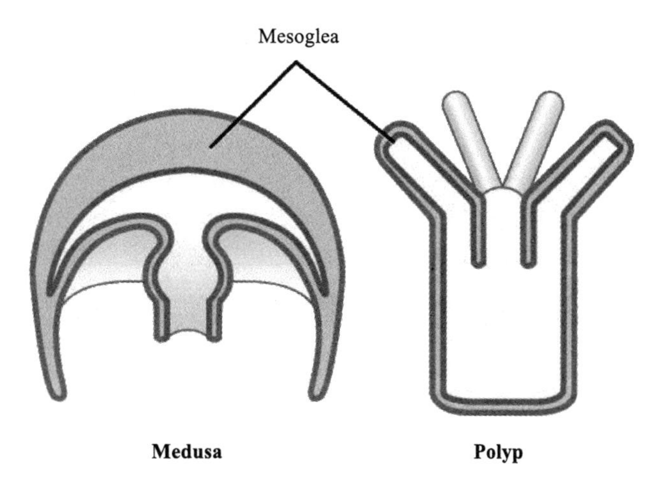

Fig. 4-1 Two body plans of cnidarians

All cnidarians show the presence of two membrane layers in the body, which are derived from the endoderm and ectoderm of the embryo. The outer layer (from ectoderm) is called the epidermis and lines the outside of the animal, whereas the inner layer (from endoderm) is called the gastrodermis and lines the digestive cavity (Fig. 4-2). Between these two membrane layers is a non-living, jelly-like connective layer—mesoglea. In terms of cellular complexity, cnidarians show the presence of differentiated cell types in each tissue layer: nerve cells, contractile epithelial cells, enzyme-secreting cells, nematocyst (Fig. 4-3), and nutrient-absorbing cells, as well as the presence of intercellular connections. However, the development of organs or organ systems is not advanced in this phylum.

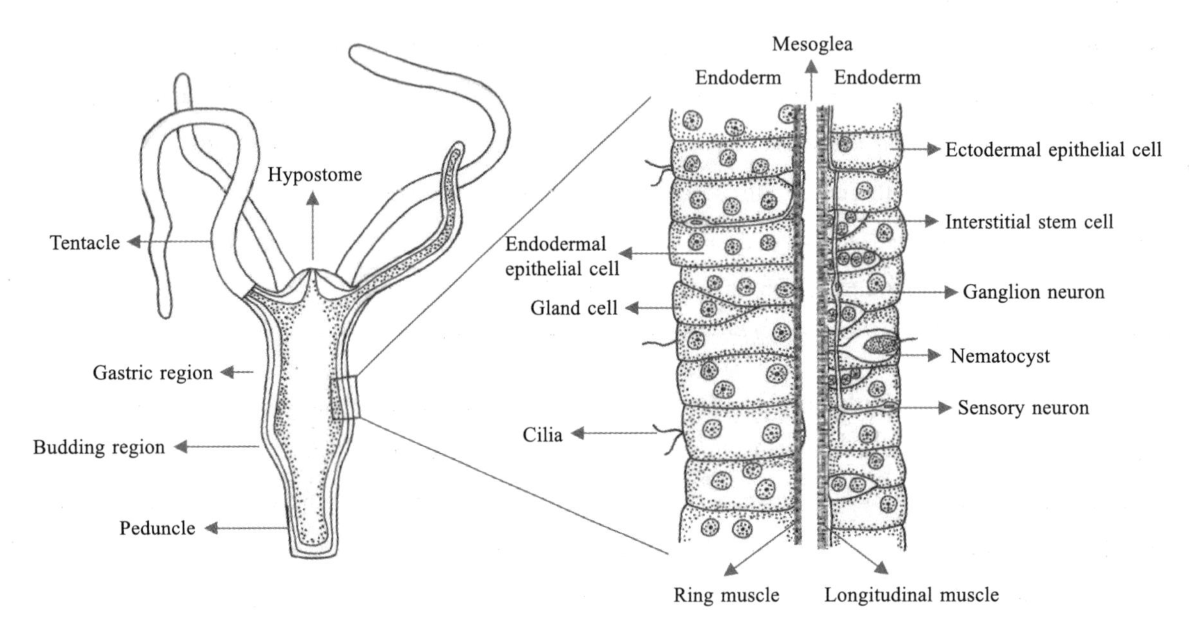

Fig. 4-2 Two membrane layers of cnidarians

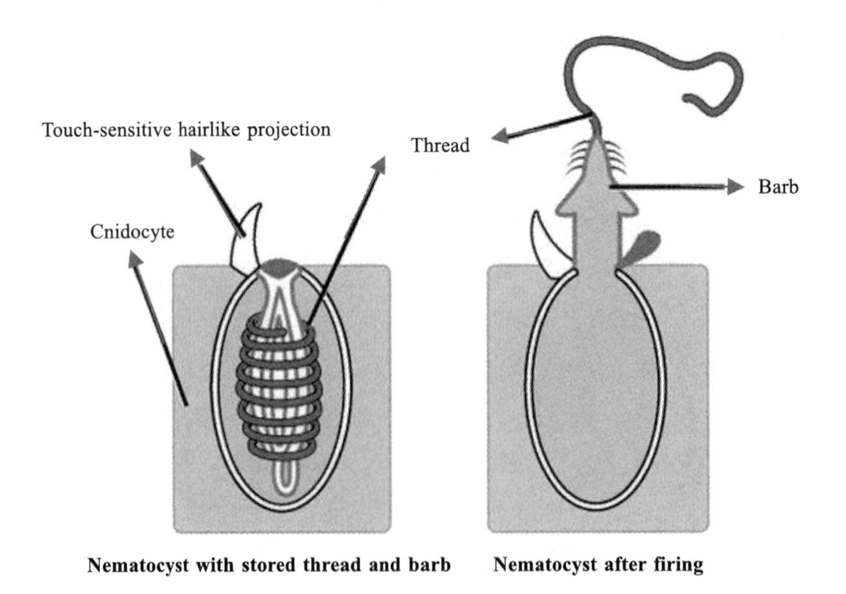

Fig. 4-3 Nematocyst

The nervous system is primitive, with nerve cells scattered across the body (Fig. 4-4). This nerve net may show the presence of groups of cells in the form of nerve plexuses or nerve cords. The nerve cells show mixed characteristics of motor as well as sensory neurons. The predominant signaling molecules in these primitive nervous systems are chemical peptides, which perform both excitatory and inhibitory functions. Despite the simplicity of the nervous system, it coordinates the movement of tentacles, the drawing of captured prey to the mouth, the digestion of food, and the expulsion of waste.

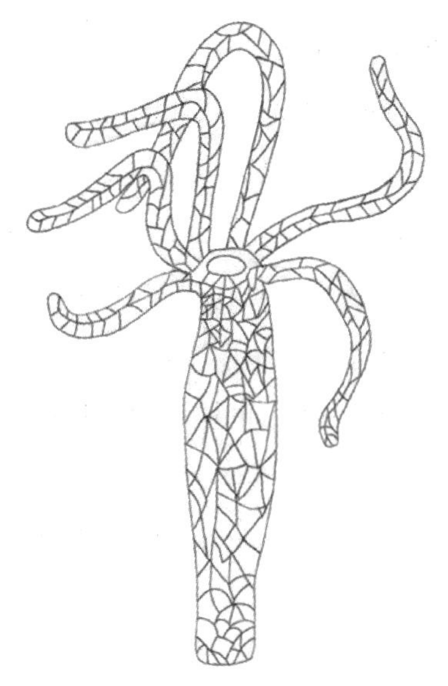

Fig. 4-4 Nervous system

The cnidarians perform extracellular digestion in which the food is taken into the gastrovascular cavity, enzymes are secreted into the cavity, and the cells lining the cavity absorb nutrients. The gastrovascular cavity has only one opening that serves as both a mouth and an anus; this is termed an incomplete digestive system. Cnidarian cells exchange oxygen and carbon dioxide by diffusion between cells in the epidermis with water in the environment, and between cells in the gastrodermis with water in the gastrovascular cavity. The lack of a circulatory system to move dissolved gases limits the thickness of the body wall, necessitating a non-living mesoglea between the layers. There is no excretory system or organs; nitrogenous wastes simply diffuse from the cells into the water outside the animal or in the gastrovascular cavity. There is also no circulatory system, so nutrients must move from the cells that absorb them in the lining of the gastrovascular cavity through the mesoglea to other cells.

● **Classification of Phylum Cnidaria**

The phylum Cnidaria contains about 10,000 described species divided into four classes: Anthozoa, Scyphozoa, Cubozoa, and Hydrozoa. The anthozoans, the sea anemones and corals, are all sessile species, whereas the scyphozoans and cubozoans are swimming forms. The hydrozoans contain sessile forms and swimming colonial forms like the Portuguese man o' war (*Physalia physalis*).

4–2 Class 1: Hydrozoa

Life histories may involve both polypoid and medusoid stages, but either may be suppressed or absent. Tetramerous or radially symmetrical medusae are small, with shelf of tissue (velum) across lower part of bell, which reduces diameter of subumbrella aperture (condition known as craspedote). Colonial forms are commonly polymorphic. Coelentera are undivided. Gametes are ripen in ectoderm. This is the only class with some freshwater members (about 2,700 species).

The polyp form in these animals often shows a cylindrical morphology with a central gastrovascular cavity lined by the gastrodermis. The gastrodermis and epidermis have a simple layer of mesoglea sandwiched between them. A mouth opening, surrounded by tentacles, is present at the oral end of the animal. Many hydrozoans form colonies that are composed of a branched colony of specialized polyps that share a gastrovascular cavity, such as in the colonial hydroid *Obelia*. Colonies may also be free-floating and contain medusoid and polypoid individuals as in the Portuguese man o' war or by-the-wind sailor (*Velella velella*). Other species are solitary polyps (*Hydra*) or solitary medusae (*Gonionemus*). The true characteristic shared by all these diverse species is that their gonads for sexual reproduction are derived from epidermal tissue, whereas in all other cnidarians they are derived from gastrodermal tissue.

Fertilization occurs either in the sea water where the germ cells are set free, or the spermatozoa may be carried by water currents to the female medusae and fertilize the ova *in situ*. Zygote forms after fertilization and immediately undergoes cleavage.

The cleavage is holoblastic and a blastula is formed. By invagination, the blastula is converted into an oval, ciliated planula larva. The planula consists of an outer layer of ciliated ectoderm and an inner mass of endoderm cells enclosing a space, the rudiment of coelenteron. The planula swims freely for a brief period and settles down on some submerged substratum by one end. The proximal end gradually narrows down and a disc appears for attachment. The distal end expands and by developing a manubrium and a circlet of tentacles, it turns to a hydrula or simple polyp. The hydrula sends out lateral buds and, by a repetition of this process, it is converted into a complex *Obelia* colony (Fig. 4-5).

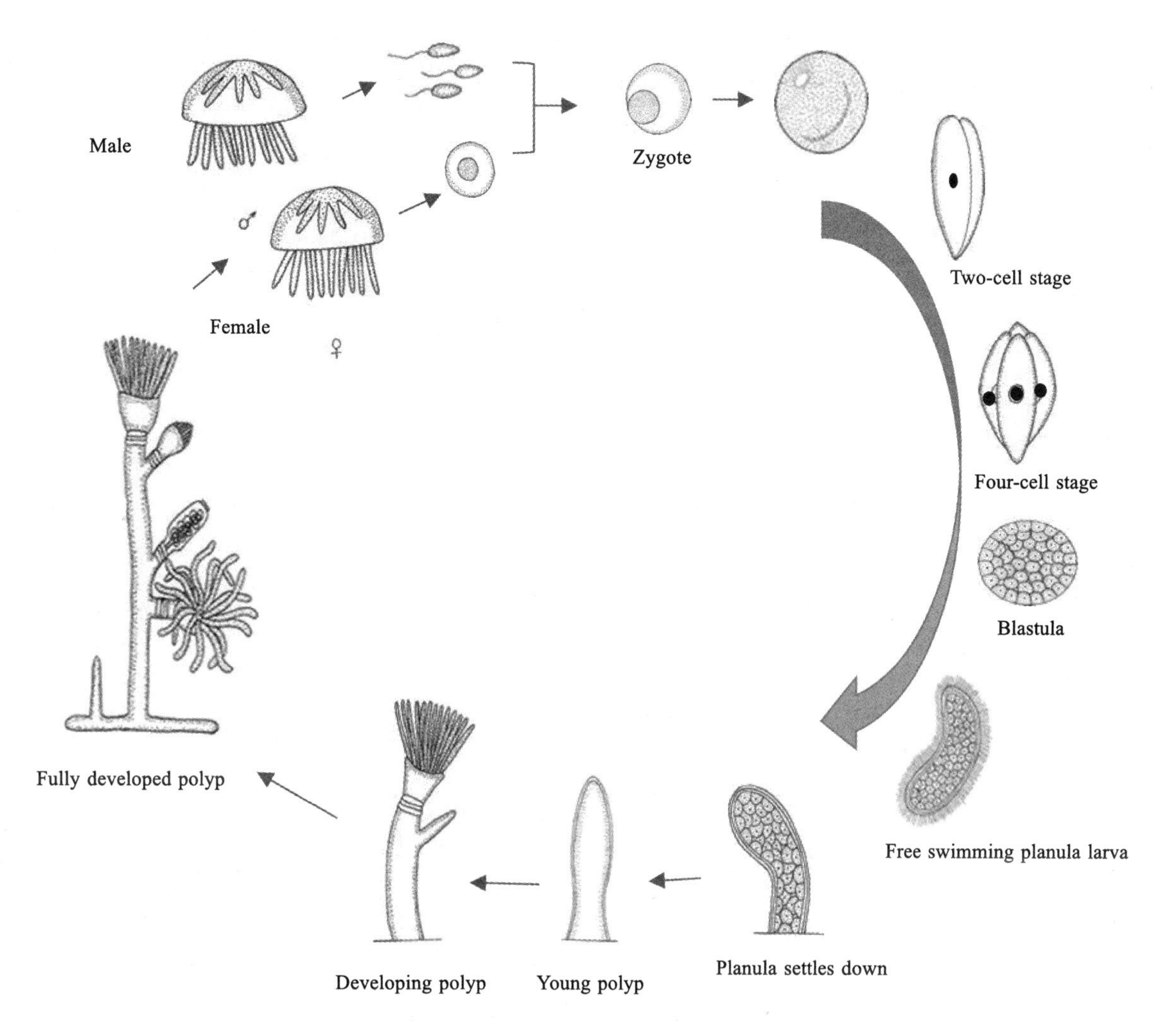

Fig. 4-5 Life cycle of *Obelia*

4–3 Class 2: Scyphozoa

Scyphozoans are exclusively marine group in which acraspedote medusae predominate. Life histories commonly involve alternation of a very small polyp, the scyphistoma, with a medusa, which develops from an ephyra released by the polyp. Coelenteron of both divided by four longitudinal septa producing tetramerous radial symmetry. Gonads are endodermal. Marginal sensory structures (rhopalia) with statocysts and/or ocelli. Most are abundant in coastal waters, but oceanic species exist. This class includes about 200 species.

Scyphozoans have a ring of muscles that lines the dome of their bodies; these structures provide them with the contractile force they need to swim through water (Fig. 4-6).

Scyphozoans have separate sexes and form planula larvae through external fertilization. Jellies exhibit the polyp form, known as a scyphistoma, after their larvae settle on a substrate; these forms will later bud-off and transform into their more prominent medusa forms.

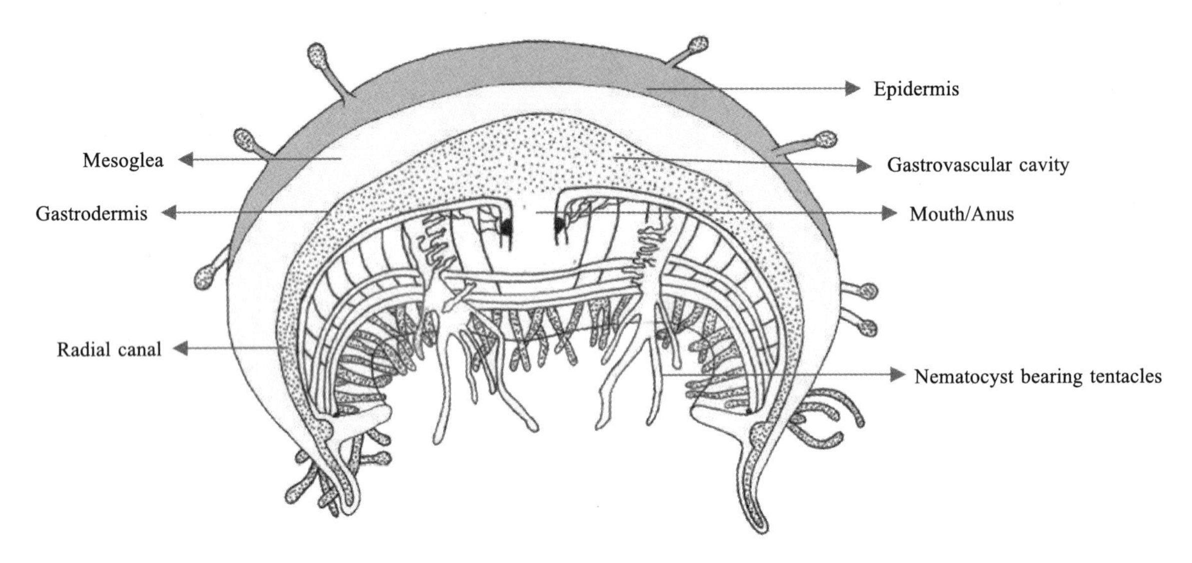

Fig. 4-6 Schematic of scyphozoans

In the jellyfish, a mouth opening, surrounded by nematocyst bearing tentacles, is present on the underside of the animal. Scyphozoans live most of their life cycle as free-swimming, solitary carnivores. The mouth leads to the gastrovascular cavity, which may be sectioned into four interconnected sacs, called diverticula. In some species, the digestive system may be further branched into radial canals. Like the septa in anthozoans, the branched gastrovascular cells serve to increase the surface area for nutrient absorption and diffusion; thus, more cells are in direct contact with the nutrients in the gastrovascular cavity.

In scyphozoans, nerve cells are scattered over the entire body. Neurons may even be present in clusters called rhopalia. These animals possess a ring of muscles lining the dome of the body, which provides the contractile force required to swim through water. Scyphozoans are dioecious animals, having separate sexes. The gonads are formed from the gastrodermis with gametes expelled through the mouth. Planula larvae are formed by external fertilization; they settle on a substratum in a polypoid form known as scyphistoma. These forms may produce additional polyps by budding or may transform into the medusoid form. The life cycle of these animals can be described as polymorphic because they exhibit both a medusal and polypoid body plan at some point.

The life cycle of a jellyfish includes two stages: the medusa stage and the polyp stage (Fig. 4-7). The polyp reproduces asexually by budding, while the medusa reproduces sexually.

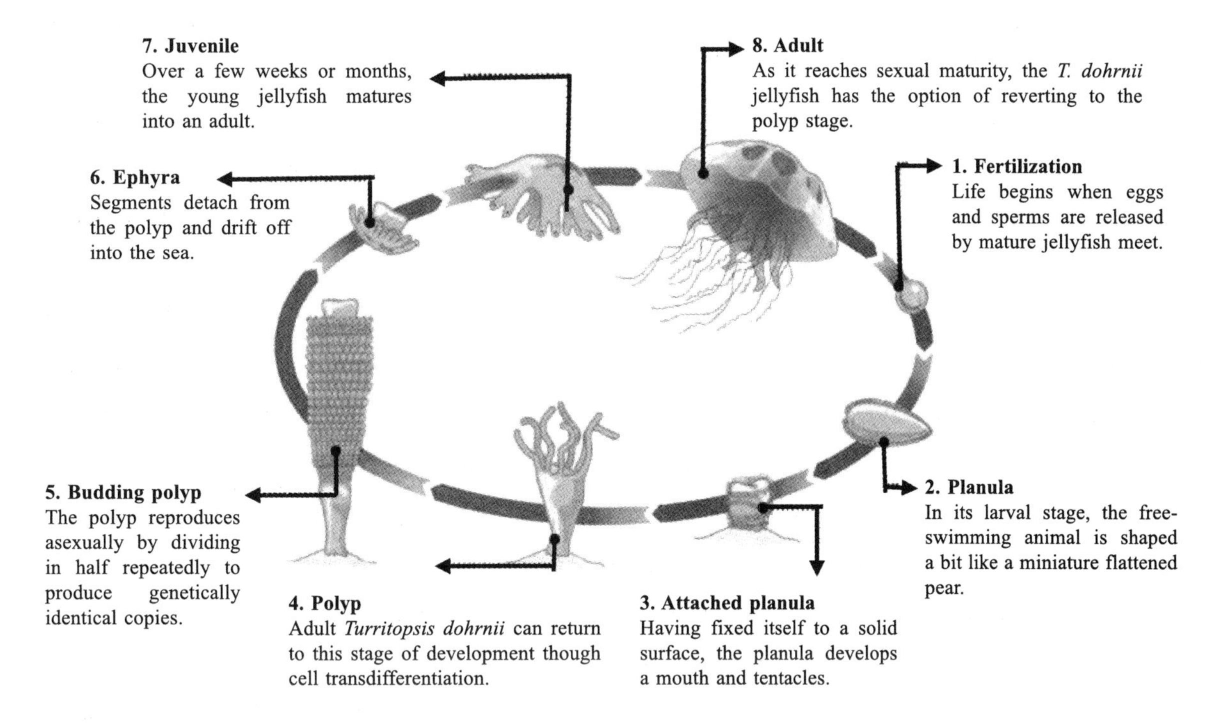

Fig. 4-7 Life cycle of a jellyfish

4-4 Class 3: Anthozoa

Anthozoans are exclusively polypoid with biradial symmetry (Fig. 4-8). Oral end seems to be a disk with central mouth and hollow tentacles arising at margin and/or on surface. Mouth leads to coelenteron via stomodaeum that has ciliated troughs (siphonoglyphs) for water transport into and out of coelenteron. Coelenteron is divided by radial mesenteries that extend inward and insert on the stomodaeum (complete mesenteries) or not (incomplete mesenteries). There are about 6,000 species. Anthozoans include sea anemones, sea pens, and corals.

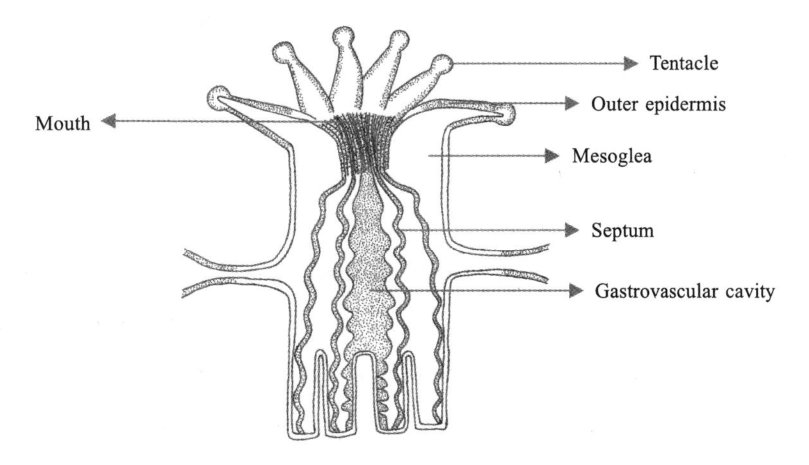

Fig. 4-8 Schematic of anthozoans

The pharynx (ingesting as well as egesting food) of anthozoans leads to the gastrovascular cavity, which is divided by mesenteries. In anthozoans, gametes are produced by the polyp; if they fuse, they will give rise to a free-swimming planula larva, which will become sessile once it finds an optimal substrate. Sea anemones and corals are examples of anthozoans that form unique mutualistic relationships with other animal species; both sea anemones and corals benefit from food availability provided by their partners.

The mouth of a sea anemone is surrounded by tentacles that bear cnidocytes. They have slit-like mouth openings and a pharynx, which is the muscular part of the digestive system that serves to ingest as well as egest food. It may extend for up to two-thirds the length of the body before opening into the gastrovascular cavity. This cavity is divided into several chambers by longitudinal septa called mesenteries. Each mesentery consists of one ectodermal and one endodermal cell layer with the mesoglea sandwiched in between. Mesenteries do not divide the gastrovascular cavity completely; the smaller cavities coalesce at the pharyngeal opening. The adaptive benefit of the mesenteries appears to be an increase in surface area for absorption of nutrients and gas exchange.

Sea anemones feed on small fish and shrimp, usually by immobilizing their prey using the cnidocytes. Some sea anemones establish a mutualistic relationship with hermit crabs by attaching to the crab's shell. In this relationship, the anemone gets food particles from prey caught by the crab, while the crab is protected from the predators by the stinging cells of the anemone. Anemone fish, or clownfish, are able to live in the anemone since they are immune to the toxins contained within the nematocysts. Another type of anthozoan that forms an important mutualistic relationship is reef building coral. These hermatypic corals rely on a symbiotic relationship with zooxanthellae. The coral gains photosynthetic capability, while the zooxanthellae benefit by using nitrogenous waste and carbon dioxide produced by the cnidarian host.

Anthozoans remain polypoid throughout their lives. They can reproduce asexually by budding or fragmentation, or sexually by producing gametes. Both gametes are produced by the polyp, which can fuse to give rise to a free-swimming planula larva. The larva settles on a suitable substratum and develops into a sessile polyp.

4–5 Class 4: Cubozoa

Class Cubozoa includes jellies that have a box-shaped medusa—a bell that is square in cross-section (Fig. 4-9); hence, they are colloquially known as "box jellyfish". These species may achieve sizes of 15–25 cm. Cubozoans display overall morphological and anatomical characteristics that are similar to those of the scyphozoans. A prominent difference between the two classes is the arrangement of tentacles. This is the most venomous group of all the cnidarians.

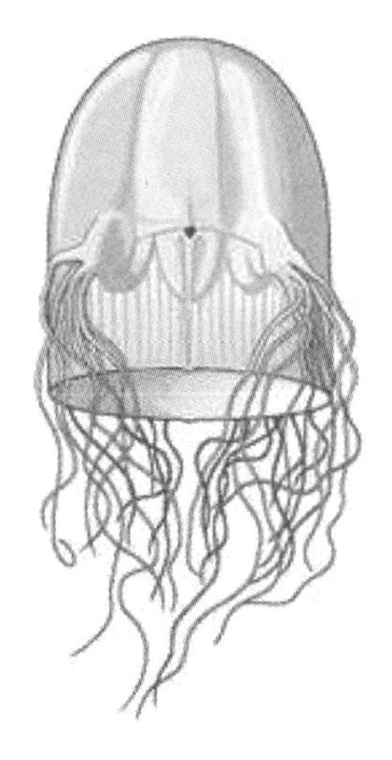

Fig. 4-9 Schematic of cubozoans

The cubozoans contain muscular pads called pedalia at the corners of the square bell canopy, with one or more tentacles attached to each pedalium. These animals are further classified into orders based on the presence of single or multiple tentacles per pedalium. In some cases, the digestive system may extend into the pedalia. Nematocysts may be arranged in a spiral configuration along the tentacles; this arrangement helps to effectively subdue and capture prey. Cubozoans exist in a polypoid form that develops from a planula larva. These polyps show limited mobility along the substratum. As with scyphozoans, they may bud to form more polyps to colonize a habitat. Polyp forms then transform into the medusoid forms.

4-6 学习要点

● 代表动物——水螅Hydrozoa的主要形态和结构特征、生态和生殖特征，刺细胞的结构和种类，皮肌细胞的形态和功能。

● 水螅纲的主要特征（体具水螅型和水母型；水螅体内无隔片；水螅水母为小型水母，具有缘膜，生殖腺由外胚层产生，多具垂管，水管简单）；薮枝螅*Obelia*的形态和结构及其世代交替现象；常见的种类（筒螅*Tubularia*、桃花水母*Craspedacusta*、钩手水母*Gonionemus*、僧帽水母*Physalia*）及其多态现象。

● 钵水母纲Scyphozoa的主要特征（水螅体退化而水母体发达，没有缘膜，胃管系统复杂，生殖腺由内胚层产生）；水螅水母与钵水母的不同点；海蜇*Rhopilema esculentum*的生活史及其世代交替现象、口腕的构造。

● 珊瑚纲Anthozoa的主要特征（只有水螅型，没有水母型；水螅体具口道、内胚层形成的隔片；生殖腺由内胚层产生）；海葵的形态和构造、其与水螅纲螅形体的不同点；本纲动物的重要类群（八放珊瑚亚纲和六放珊瑚亚纲）；我国沿海常见的种类（海鸡冠*Alcyonium*、海鳃*Pennatula*、笙

珊瑚 *Tubipora*、红珊瑚 *Corallium*、鹿角珊瑚 *Madrepora*、脑珊瑚 *Meandrina* 等）；珊瑚骨骼的形成及其在自然地理上的作用。

- 水螅纲水螅体与珊瑚纲水螅体的区别，水螅纲水母与钵水母纲水母的区别。

4–7 巩固测验

【名词】

消化循环腔、浮浪幼虫、刺细胞、世代交替

【选择】

1. 腔肠动物具有（　　）。

A. 骨针细胞　B. 皮肌细胞　C. 焰细胞　D. 领细胞

2. 海产腔肠动物在发育过程中需经历（　　）。

A. 担轮幼虫期　B. 两囊幼虫期　C. 牟勒氏幼虫期　D. 浮浪幼虫期

3. 腔肠动物中，生殖腺不是起源于外胚层的是（　　）。

A. 水螅　B. 钩手水母　C. 海葵　D. 桃花水母

4. 下列动物中具有典型的世代交替现象的是（　　）。

A. 海葵　B. 水螅　C. 薮枝虫　D. 珊瑚

5. 腔肠动物所特有的结构是（　　）。

A. 身体次生性辐射对称　B. 具完全消化道　C. 具有刺细胞　D. 具有链状神经系

6. 海葵的身体为（　　）。

A. 次生性辐射对称　B. 辐射对称　C. 两侧对称　D. 两辐射对称

7. 水螅纲区别于钵水母纲的特征是（　　）。

A. 生活史有世代交替　B. 生殖腺来源于外胚层

C. 具有水螅型和水母型　D. 具有刺细胞

【简答】

1. 腔肠动物主要特征有哪些？
2. 薮枝螅和海蜇的生活史各是怎样的？
3. 绘图标注水螅体壁的细胞类型。

Chapter 5 Phylum Platyhelminthes

5-1 Characteristics of Phylum Platyhelminthes

- Their body is dorsoventrally flattened.
- They exhibit bilateral symmetry.
- They are triploblastic, with three germ layers.
- They do not have a body cavity and are acoelomate.
- Body is soft and unsegmented.
- They are mostly parasitic with a few free-living.
- They exhibit an organ system grade of organization.
- The digestive system is incomplete or absent. There is a single opening which leads to a well-developed gastrovascular cavity. The anus is absent. There is no true stomach structure. In a few species, the digestive system is completely absent.
- Respiratory and circulatory systems are absent. Respiration generally occurs through diffusion through the general body surface.
- The excretory system has protonephridia with flame cells.
- There is a primitive nervous system present.
- These animals are hermaphrodites.
- Sexual reproduction happens through gametic fusion.
- Asexual reproduction also happens in a few species through regeneration and fission.
- Fertilization is internal.
- The life cycle of these organisms can be complex, especially if they are parasitic, as this may involve one or more host animals.
- The different classes under this phylum are Turbellaria, Trematoda, and Cestoda.

Platyhelminthes are very commonly known as flatworms or tapeworms; these animals are soft-bodied invertebrate animals. There are around 20,000 species of these animals. A few of these live as parasites in humans and other animals. It is because of this parasitic nature that they do cause some amount of trouble for the host animal. A few species belonging to this phylum can be a major cause of certain diseases. Schistosomiasis, or bilharzia or bilharziasis, is a disease caused by these parasitic flatworms belonging to the family Schistosomatidae.

The most distinguishing feature of these invertebrates is their flat body. As the body does not have any cavity, they are flat. The body is also not segmented and they do not have specialized systems (Fig. 5-1). Around 80% of the flatworms are parasitic in nature, while a few free-form flatworms are also present. The free-living species are scavengers or predators. The parasitic species feed on the tissues of the host organism in which they live.

The animals in this phylum have a diverse range in size. Some are microscopic, while a few go up to two feet (about 0.6 m) long. They are also hermaphrodites, which mean that both the sexes are present in the same organism.

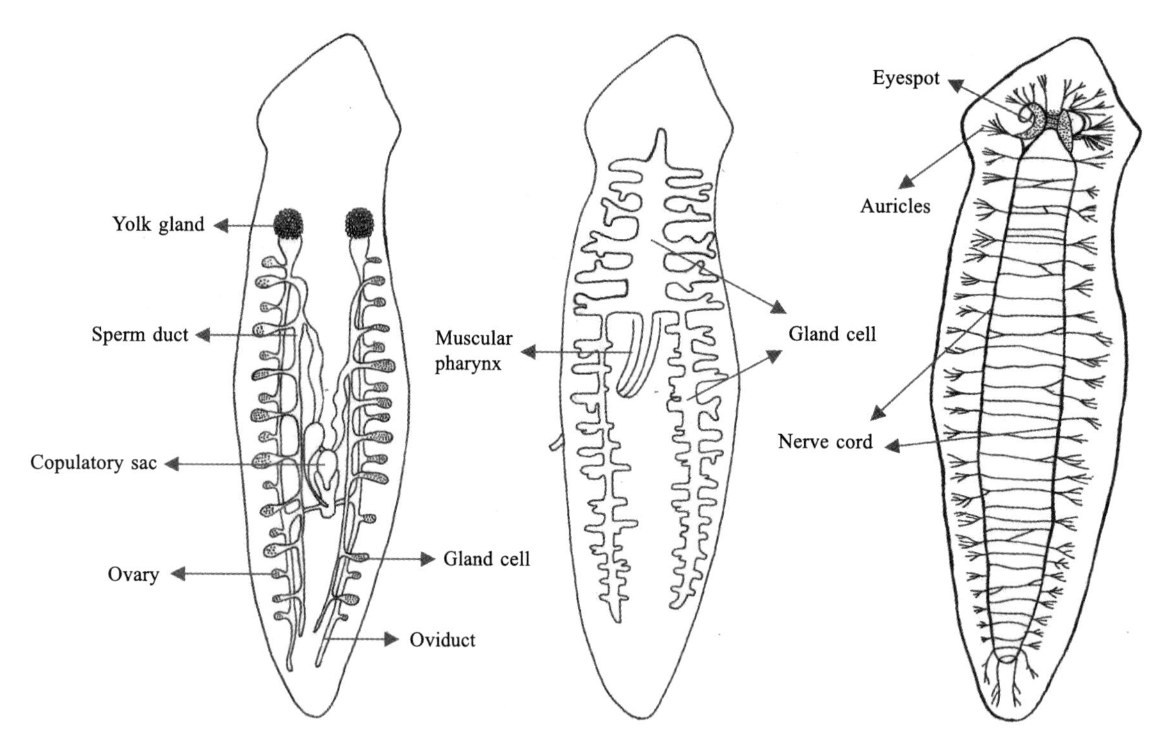

Fig. 5-1 Schematic of planarians

5-2 Class 1: Turbellaria

● **Example: *Dugesia japonica***

The mostly free-living, primarily carnivorous, flatworms of class Turbellaria are characterized by a soft epidermis that is ciliated, at least on the ventral surface. The movement of the cilia propels the smaller forms. Larger species glide along by muscular waves, usually over mucous beds secreted by special cells.

The planarian is a flatworm that has a gastrovascular cavity with one opening that serves as both mouth and anus. The excretory system is made up of tubules connected to excretory pores on both sides of the body. The nervous system is composed of two interconnected nerve cords running the length of the body, with cerebral ganglia and eyespots at the anterior end (Fig. 5-2).

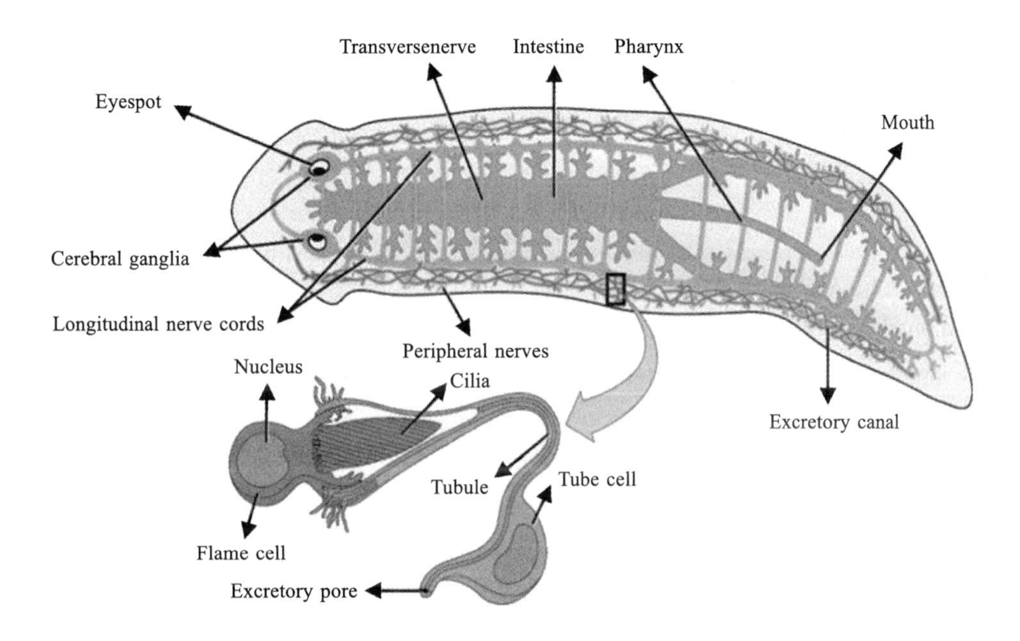

Fig. 5-2 Schematic of *Dugesia japonica*

5–3 Class 2: Trematoda

● **Example: *Clonorchis sinensis***

The parasitic flatworms of class Trematoda, also called flukes, have oral suckers, sometimes supplemented by hooks, with which they attach to their vertebrate hosts (Fig. 5-3). Trematodes have retained the same body form and digestive cavity as the turbellarians. However, practically the entire interior is occupied by the reproductive system; the organism is capable of producing huge numbers of offspring. Trematodes of the order Digenea have complex life cycles involving two or more hosts. The larval worms occupy small animals, typically snails and fish, and the adult worms are internal parasites of vertebrates. Many species, such as the liver fluke *Clonorchis sinensis* and the blood fluke (*Schistosoma*), cause serious diseases in humans.

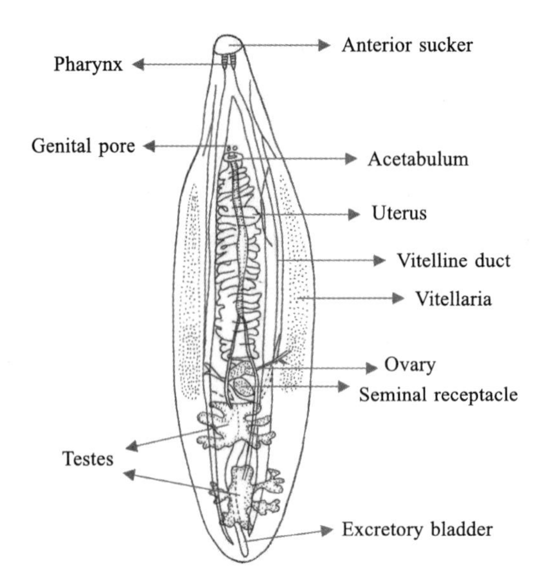

Fig. 5-3 Schematic of *Clonorchis sinensis*

The life cycle of *Clonorchis sinensis* starts with a free-floating egg in freshwater. The egg is consumed by a snail and once inside the snail it hatches and matures. The fluke then burrows out of the snail and into the water, where it can seek out a new host. Freshwater fish are preferred hosts for the parasites, which burrow into the body before encapsulating themselves. This capsule is a critical part of the fluke's survival strategy (Fig. 5-4).

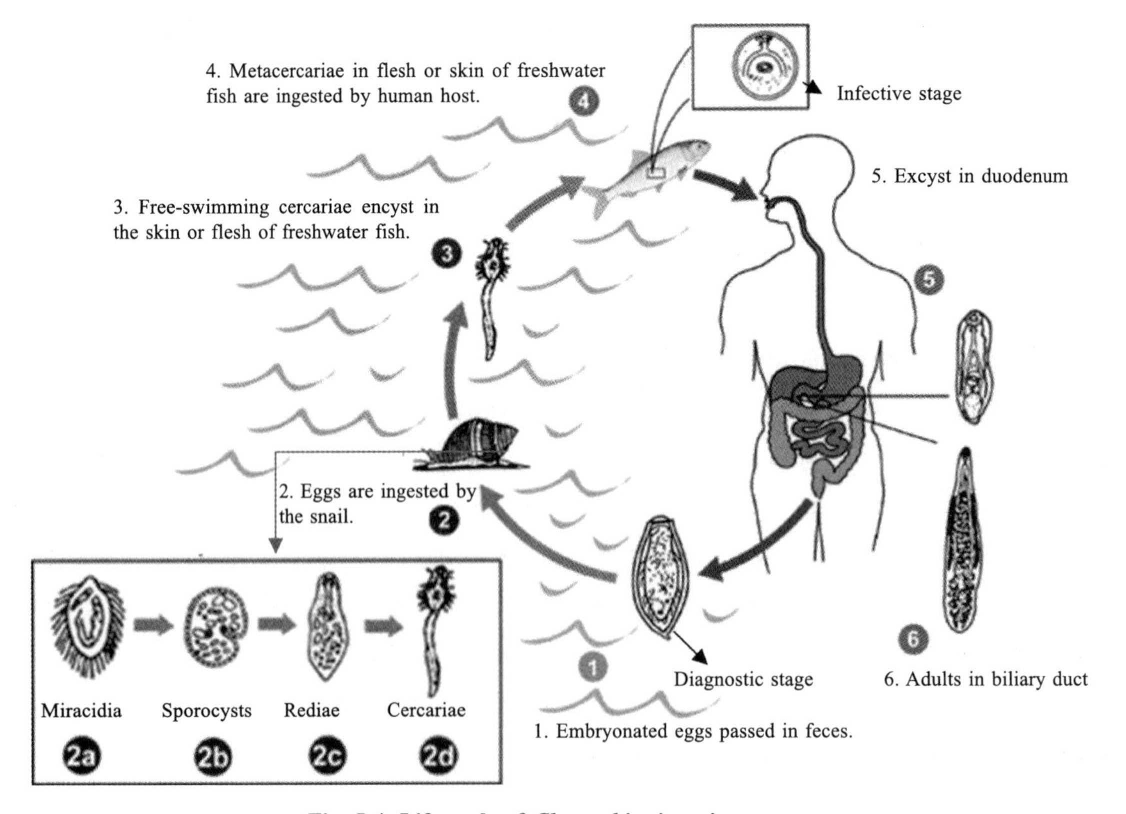

Fig. 5-4 Life cycle of *Clonorchis sinensis*

5-4 Class 3: Cestoda

● **Example: *Taenia solium***

The body of the cestodes, also known as tapeworms, has lost the typical turbellarian form. Although there are a few unsegmented species, the bulk of a typical cestode body consists of a series of linearly arranged reproductive segments called proglottids. There is no mouth or digestive system; food is absorbed through the cuticle. Adults live in the digestive tract of vertebrates, and larval forms encyst in the flesh of various vertebrates and invertebrates.

The body of *Taenia solium* is long, dorsoventrally flattened, narrow, ribbon-like, reaching a length of 2–3 m. The colour of the body is opaque-white. Body consists of scolex, neck and strobila or body segments. The anterior end of the body of *Taenia* has a knob-like scolex. The scolex is smaller than the head of a pin, about 1 mm in diameter with four cup-like muscular suckers having radial muscles, and an anterior round prominence. The rostellum has 22–32 curved, chitinous hooks in two circles, the inner circle with larger hooks and outer circle with smaller ones (Fig. 5-5).

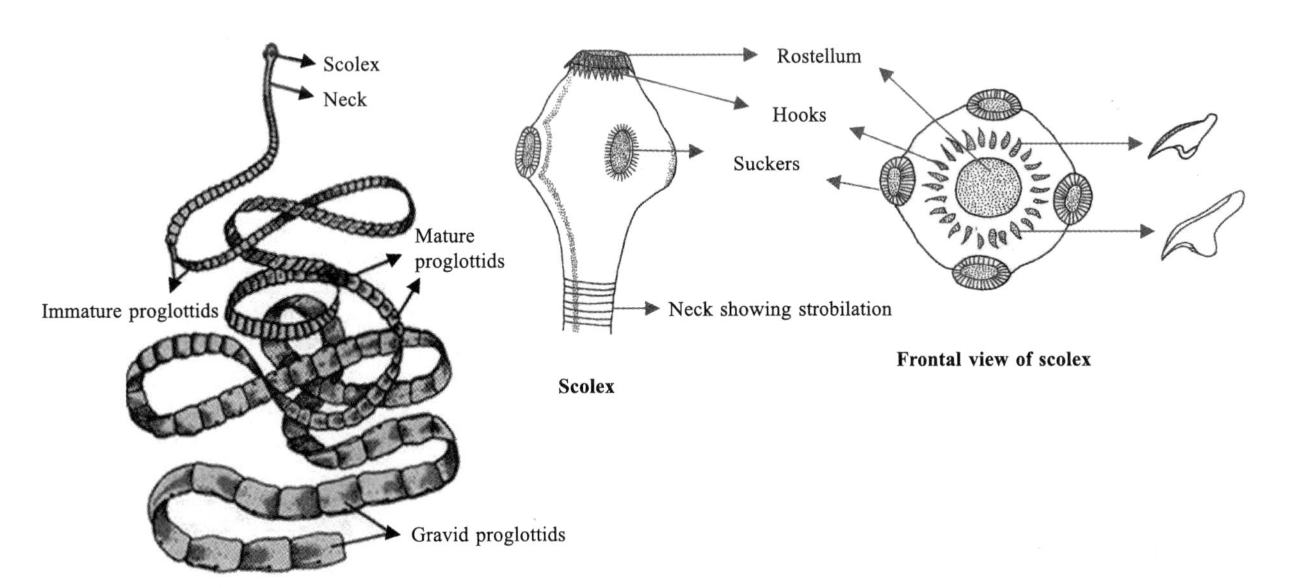

Fig. 5-5 Schematic of *Taenia sodium*

Behind the scolex is a thin, small, narrow, unsegmented neck which grows continuously and proliferates proglottids by transverse fission or asexual budding. Therefore, this region is variously called like growth zone, area of proliferation or budding zone, etc. The neck is followed by the flattened, ribbon-like body called strobila. The strobila forms the main bulk of the body and consists of a series of proglottids arranged in a linear fashion.

The body of an adult tapeworm is virtually a reproductive factory. Behind a small securing knob, called a scolex, which bears a circle of hooks or other attachment organs, the proglottids constantly bud off and gradually enlarge. As they mature, they become filled with male and female reproductive organs. Cross-fertilization takes place with adjacent worms or neighboring proglottids; in some cases, self-fertilization occurs. In some species, the ripe proglottids filled with eggs are shed. In others, the fertilized eggs leave the adult host in the feces. If the eggs are consumed by the intermediate host, the life cycle continues. Tapeworm species that infest human intestines as adults include *Taenia saginata*, *T. solium*, the dwarf tapeworm *Hymenolepsis nana*, and the fish tapeworm *Diphyllobothrium latum*, which can reach lengths of up to 50 feet (about 15 m) (Fig. 5-6).

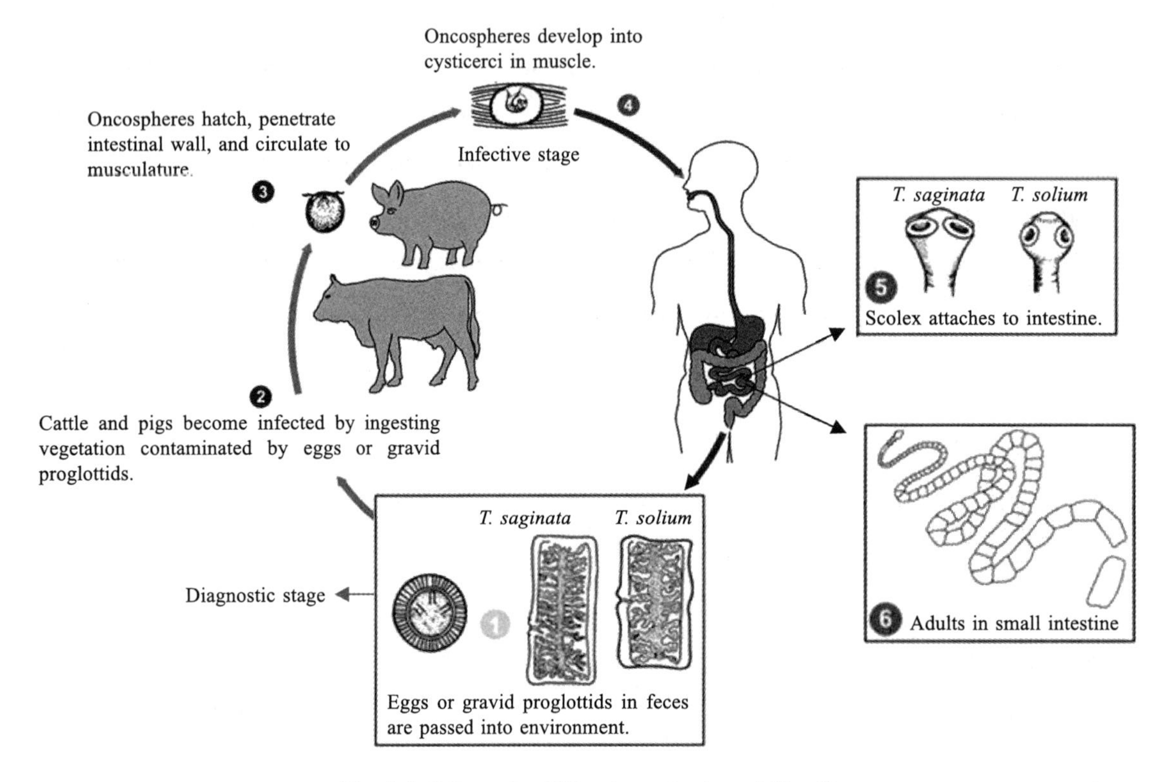

Fig. 5-6 Life cycle of *Taenia saginata* and *T. solium*

Female reproductive organs consist of a bilobed ovary, oviduct, ootype, vagina, uterus, Mehlis gland and vitelline gland. It is bilobed situated ventrally in the posterior part of the proglottid, and also called germarium. Both the lobes are connected by a transverse ovarian bridge or isthmus. Each lobe of ovary is formed of a number of radially arranged follicles. Nearly from the middle of ovarian bridge, a median short but wide oviduct arises which opens into the ootype. Ootype is a small, spherical bulb-like structure situated at the junction of oviduct, uterus and vitelline duct. Uterus is a blind club-shaped, sac-like structure; it originates from the ootype and extends towards the anterior side of the proglottid. The fertilized eggs are packed in it and in gravid proglottids, it becomes highly branched having 7–10 branches (Fig. 5-7).

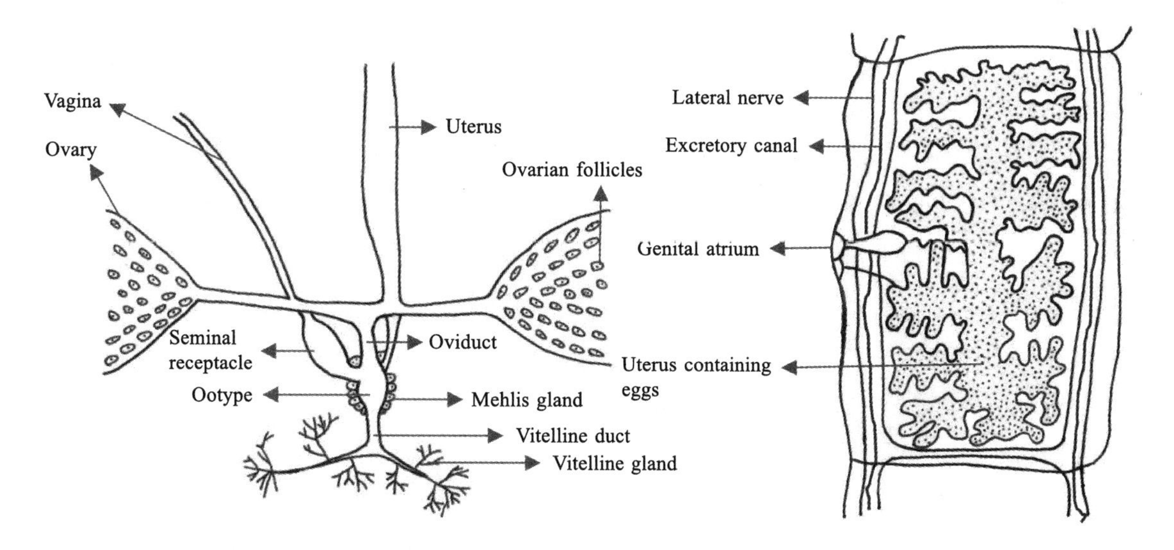

Fig. 5-7 Female reproductive organs

5-5 学习要点

- 中胚层在动物系统发生中首次出现的意义及其对动物身体机能适应环境所产生的深刻影响。
- 涡虫纲代表动物——涡虫 *Dugesia japonica* 的主要形态和结构特征。
- 吸虫纲代表动物——华支睾吸虫 *Clonorchis sinensis* 的主要形态和特点，以及其形态、结构与其寄生生活的适应关系；华支睾吸虫的生活史，其幼体生殖和更换寄主的生物学意义；吸虫纲的重要类群（单殖亚纲和复殖亚纲）；三代虫 *Gyrodactylus*、指环虫 *Dactylogyrus*、肝片吸虫 *Fasciola hepatica*、布氏姜片吸虫 *Fasciolopsis buski*、日本血吸虫 *Schistosoma japonicum* 的分类地位及其对人类的危害，它们生活史中的终末寄主、中间寄主和寄生部位；多胚生殖和卵胎生的特点。
- 中胚层的发生及其在动物系统发生中的意义。
- 寄生虫更换寄主的生物学意义。

5-6 巩固测验

【名词】

皮肌囊、合胞体、焰细胞、杆状体、原肾管

【选择填空】

1. 从扁形动物开始的以后各门动物（除棘皮动物和腹足纲外）的对称形式是（　　）。

A. 无固定的对称形式　　B. 辐射对称　　C. 左右对称　　D. 两辐射对称

2. 下列动物中具有原肾型排泄系统的是（　　）。

A. 水螅　　B. 涡虫　　C. 蚯蚓　　D. 河蚌

3. 真涡虫的神经系为（　　）。

A. 网状神经系　　B. 链状神经系　　C. H形神经系　　D. 梯状神经系

4. 涡虫耳突的功能是（　　）。

A. 听觉　　B. 感光　　C. 触觉　　D. 味觉和嗅觉

5. 其成虫是人体寄生吸虫最大的是（　　）。

A. 日本血吸虫　　B. 华支睾吸虫　　C. 姜片虫　　D. 肺吸虫

6. 猪带绦虫的卵被寄主吞食后孵出的幼虫叫（　　）。

A. 毛蚴　　B. 胞蚴　　C. 囊尾蚴　　D. 六钩蚴

7. 下列非昆虫传播的寄生虫是（　　）。

A. 日本血吸虫　　B. 疟原虫　　C. 杜氏利什曼原虫　　D. 丝虫

8. 不经血液循环途径即可到达其正常寄生部位发育为成虫的寄生虫是（　　）。

A. 日本血吸虫　　B. 绦虫　　C. 蛔虫　　D. 钩虫

9. 下列不进行宿主更换的寄生虫是（　　）。

A. 黑热病原虫　　B. 日本血吸虫　　C. 丝虫　　D. 钩虫

10. 下列寄生虫中由昆虫传播的是（　　）。

A. 羊肝蛭　　B. 猪带绦虫　　C. 血吸虫　　D. 利什曼原虫

【简答题】

1. 扁形动物的主要特征有哪些？
2. 中胚层的形成意义是什么？
3. 两侧对称体形出现的生物学意义是什么？
4. 什么是焰细胞？什么是原肾排泄系统？
5. 华支睾吸虫、日本血吸虫、猪带绦虫的生活史分别是怎样的？

Chapter 6 Pseudocoelomata

A pseudocoelomate is an organism with body cavity that is not derived from the mesoderm, as in a true coelom, or body cavity. A pseudocoelomate is also known as a blastocoelomate, as the body cavity is derived from the blastocoel, or cavity within the embryo (Fig. 6-1). A true coelom is lined with a peritoneum which serves to separate the fluid from the body cavity. In a pseudocoelomate, the body fluids bath the organs, and pseudocoelomates receive their nutrients and oxygen from the fluid in the cavity.

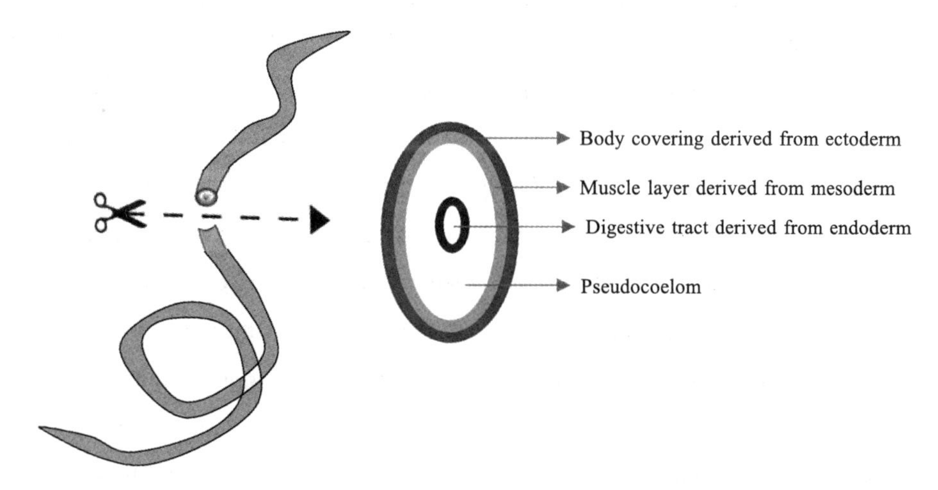

Fig. 6-1 Body cavity of pseudocoelomates

The presence of the blastocoel in the embryo is a condition universal to all metazoans. In most metazoans, the mesoderm becomes the lining of the body cavity, creating the true coelom. Some pseudocoelomates represent the primitive form of coelomates, and their ancestors never had a true coelom. Other organisms have lost the peritoneum and have regressed to the pseudocoelomate condition. The larval form of some coelomates starts as pseudocoelomates. Pseudocoelomates are often small animals, which rely mostly on diffusion to distribute oxygen and nutrients to their cells. These organisms typically have no circulatory system, or have an open circulatory system which circulates a blood-like substance known as hemolymph within the body cavities. Because of this, the cavity is referred to as the hemocoel and the organism as a hemocoelomate.

The pseudocoelomates include the nematodes, rotifers, gastrotrichs, and introverts. Some members of some other phyla are also, strictly speaking, pseudocoelomates.

6-1 Phylum Nematoda

- Body is bilaterally symmetrical and vermiform.
- Body has more than two cell layers, tissues and organs.
- Body cavity is a pseudocoel, and body fluid is under high pressure.
- Body possesses a through gut with a subterminal anus.
- Body is covered in a complex cuticle.
- The animal has a nervous system with pharyngeal nerve ring.
- The animal has no circulatory system (no blood system).
- Reproduction is normally sexual and gonochoristic.
- The animal feeds on just about everything.
- The animal lives just about everywhere, and many species are endoparasites.

Nematodes are the most species phylum after the arthropods. They occur in nearly every habitat including as parasites in all sorts of plants and animals (they do not like dry places however). One species (*Turbatrix aceti*) is known that can live in old vinegar and another that has only been found in German beer mats. Though only about 80,000 species have been described, some scientists estimate there may be as many as one million species all told. They can occur in very dense numbers in the soil and rotting vegetation. As many as 90,000 have been found in a single rotting apple, while millions occur in the top 3 cm of a square metre of good quality soil. While there are a huge number of free-living nematodes, there are also a large number of parasitic species, many of which cause diseases to man and other animals as well as to plants. Nearly every living organism has been found to be parasitized by one species of nematode or another. Most nematodes are reasonably small. They range in size from 100 μm in length to the female giant nematode *Dioctophyme renale* which may be up to 1 m long.

Human beings, along with all other living things are host to numerous nematode parasites. The most common of these is *Ascaris lumbricoides* with an estimated 700 million people effected globally. This nematode is not normally fatal and in low numbers may have very little effect on adults, however in heavy doses it can be quite debilitating, especially for children. The nematodes infecting mankind include several species of filarial worms. The most important of these are *Wuchereria bancrofti* and *Brugia malayi* which are very similar and cause lymphatic filariasis, *Onchocerca volvulus* which causes river blindness and *Loa loa* which causes loiasis. Other species are *Dracunculus medinensis* known as Guinea worm, *Trichinella spiralis* causing trichinosis, *Necator americanus* and *Ancylostoma duodenale* causing hookworm, *Enterobius vermicularis* causing pinworm and *Trichuris trichiura* causing whipworm or trichuriasis.

● **Example:** ***Ascaris lumbricoides***

It is a common endoparasite in the intestine of human beings. Body is worm-like and has pseudo-segmentation. The females are larger and stouter than males and measure 20–30 cm in length, whereas males reach to 15–26 cm in length (Fig. 6-2). Mouth has three distinct lips. Pharynx has few caeca but no posterior. The posterior end of female is straight and that of male is curved. Malc bcars a pair of sub-terminal Denial setae at posterior end. Body is marked with four longitudinal canals or lines which run along the whole body length (Fig. 6-3). Excretory pore lies near the anterior end and the female genital opening lies near the middle. The fertilized eggs pass out with feces and hatch into embryos in warm and moist soil and infection occurs through contaminated hands and drinking water.

Adult worms live in the lumen of the small intestine. A female may produce approximately 200,000 eggs per day, which are passed with the feces. Unfertilized eggs may be ingested but are not infective. Larvae develop to infectivity within fertile eggs after 18 days to several weeks, depending on the environmental conditions (optimum: moist, warm, shaded soil). After infective eggs are swallowed, the larvae hatch, invade the intestinal mucosa, and are carried via the portal, then systemic circulation to the lungs. The larvae mature further in the lungs (10–14 days), penetrate the alveolar walls, ascend the bronchial tree to the throat, and are swallowed. Upon reaching the small intestine, they develop into adult worms (Fig. 6-4). Between two and three months are required from ingestion of the infective eggs to oviposition by the adult female. Adult worms can live 1–2 years.

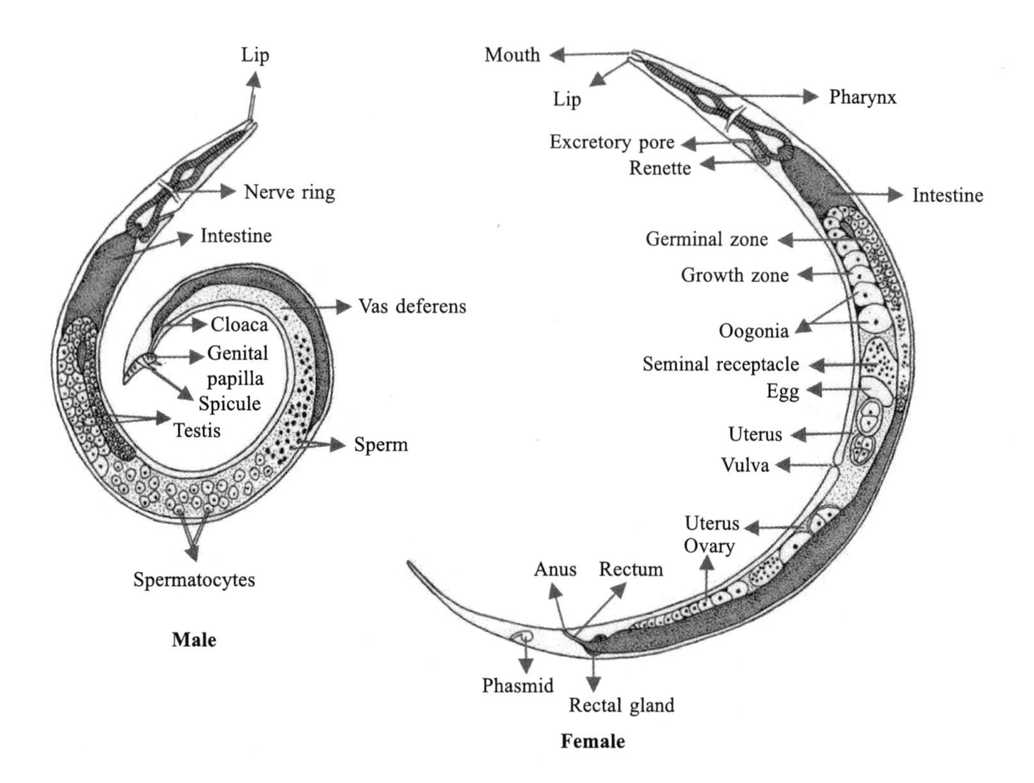

Fig. 6-2 Male and female *Ascaris lumbricoides*

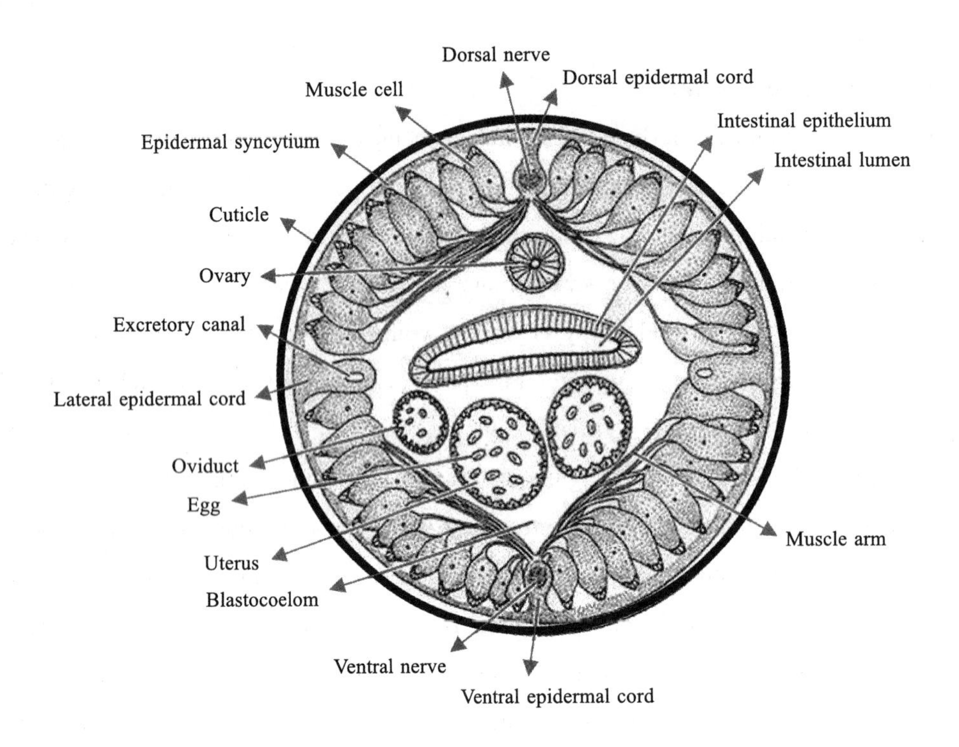

Fig. 6-3 Cross-section view of *Ascaris lumbricoides*

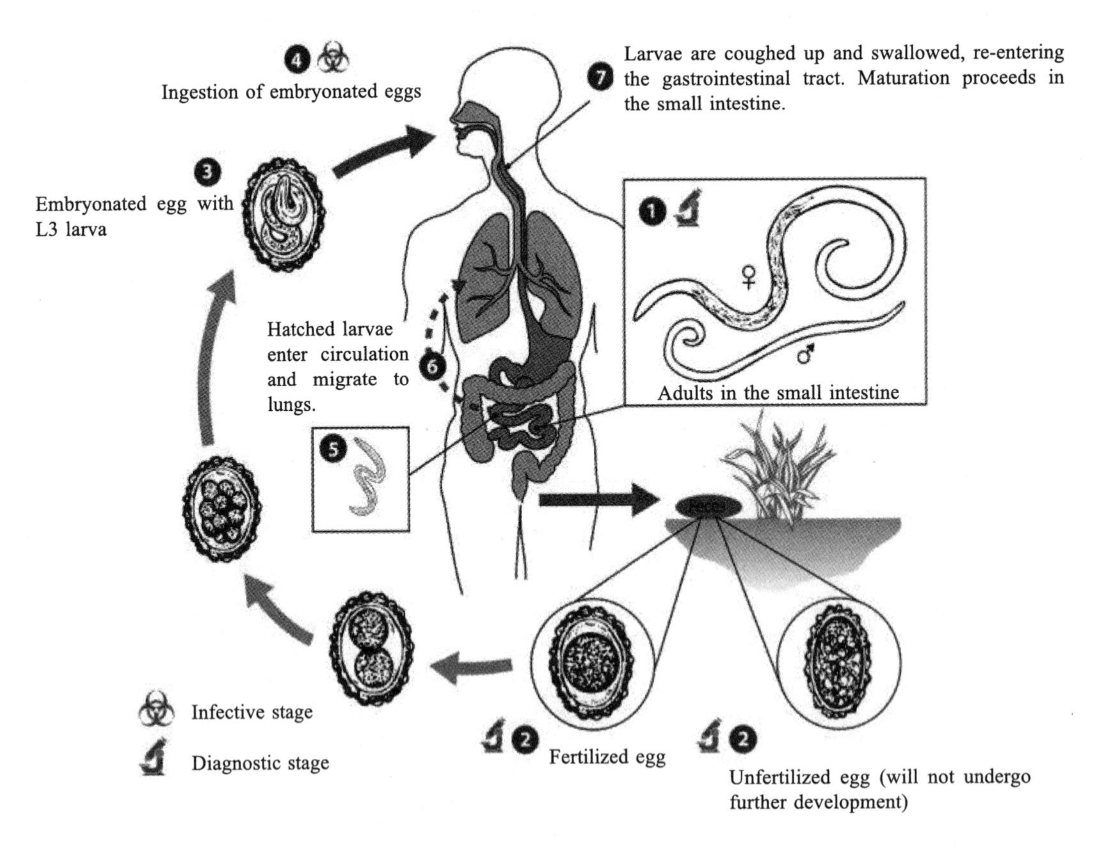

Fig. 6-4 Life cycle of *Ascaris lumbricoides*

6–2 Phylum Rotifera

- Body is bilaterally symmetrical.
- Body has more than two cell layers, tissues and organs.
- Body cavity is a pseudocoelom.
- Body possesses a through gut with an anus.
- Body is covered in an external layer of chitin called a lorica.
- The animal has a nervous system with a brain and paired nerves.
- The animal has no circulatory or respiratory organs.
- Reproduction is mostly parthenogenetic, otherwise sexual and gonochoristic.
- The animal feeds on bacteria and protista, or is parasitic.
- All live in aquatic environments either free swimming or attached.

Rotifers are small, translucent or transparent, cylindrical animals which all have a ring of cilia around the head—the name "rotifer" derives from this feature (Fig. 6-5). Rotifers range in size from 0.1 mm to 1 mm long, but most are capable of considerable longitudinal and radial expansion or contraction. Rotifers constitute a well-defined phylum quite unlike any other animal, but their

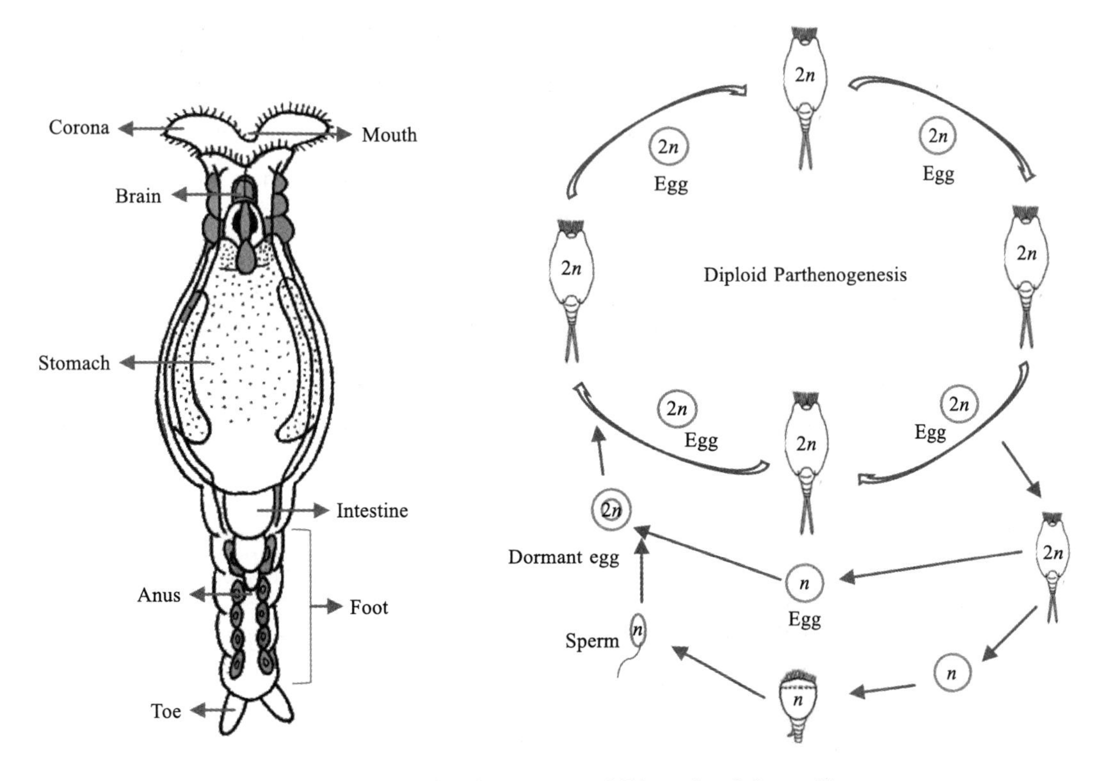

Fig. 6-5 Structure and life cycle of the rotifer

evolutionary relationship to other phyla remains unclear: they have been thought related to Nematoda, Nematomorpha and other small phyla in a group known as Aschelminthes, but some evidence points to a relationship with acanthocephalans. There are an estimated 2,000 species known worldwide, many cosmopolitan, but there are undoubtedly many undescribed species.

Rotifers are either suspension feeders (Digononta), predators on protozoa, other rotifers and other small metazoans (Monogononta), or herbivores on filamentous algae (Digononta & Monogononta). The ring of cilia (known as the corona) and the pharynx (known as the mastax), are modified to adapt to these different modes of feeding. A few species are endoparasites of snail eggs, the intestine or body cavity of earthworms, freshwater oligochaetes and molluscs. One genus lives within the filaments of a freshwater alga. Rotifers are found in huge numbers in most freshwater habitats, but they also occur in moist terrestrial habitats, particularly moss beds. A few species occur in estuarine or marine habitats, and a few are parasitic. Most rotifers are solitary, but colonial species are known.

Rotifers have a short, direct, viviparous life cycle (Fig. 6-5). Reproduction is by parthenogenesis in most species, while sexual reproduction with separate sexes also occurs in some groups. In parthenogenetic species, eggs are produced by mitosis and always hatch into females. Females of mobile species hatch as small versions of adults and achieve sexual maturity after a short period. Sessile species may have a very short planktonic stage, which resembles free-swimming rotifers, before setting and assuming adult shape. In species reproducing sexually, parthenogenesis is still the most common method of reproduction for most of the time. Sexual reproduction only takes place at certain times, and is induced by environmental factors such as high abundance, changes in food supply, photoperiod or temperature. When males occur, they are always smaller than females, hatch sexually mature, lack a functional gut and have no growth period before mating and dying after a very short time. Males are haploid and are produced following meiosis. Most rotifers have no true cuticle and do not moult.

The life cycle is typically a few days to two weeks, but eggs may be produced as little as two days after the parent itself hatched. The maximum life span is about two months. Different types of eggs are produced. Some have thin shells and hatch within a few days. Others may form a resistant resting stage capable of survival for many years before hatching. Females usually lay 10–50 eggs.

6–3 学习要点

● 原肾排泄系统中腺型和管型的区别；假体腔在动物系统发生中出现的意义及其对动物身体机能所产生的深刻影响。

● 线虫动物门代表动物——人蛔虫*Ascaris lumbricoides*的主要形态和结构特征（皮肤肌肉囊、合胞体、侧线、背线、腹线的组织学特征，筒状的神经系统）、生活史特点及对人类的危害、分布和防治原则；重要的寄生线虫（蛲虫*Enterobius vermicularis*、十二指肠钩虫*Ancylostoma duodenale*、丝虫*Wuchereria*等）及它们在人体中的寄生部位。

- 轮虫Rotifer的生活史及其有性生殖和孤雌生殖的特点。
- 假体腔的发生及其特征。

6-4 巩固测验

【名词】

假体腔、交合刺、孤雌生殖

【选择】

1. 以下不属于线虫动物门的是（　　）。

A. 秀丽线虫　　B. 钩虫　　C. 线虫　　D. 铁线虫

2. 从蛔虫围咽神经环向体后发出（　　）条神经干。

A. 2　　B. 4　　C. 6　　D. 8

【填空】

1. 假体腔动物的消化管分化出（　　）、（　　）、（　　）三段，其中（　　）来源于外胚层。

2. 蛔虫的排泄器官为（　　）型排泄器官，分布于（　　）中。

3. 蛔虫的上皮为（　　）结构。

【简答题】

线虫、轮虫动物的主要特征有哪些？轮虫、蛔虫的生活史各是怎样的？

Chapter 7 Phylum Annelida

7-1 Characteristics of Phylum Annelida

- Body is bilaterally symmetrical and vermiform.
- Body has more than two cell layers, tissues and organs.
- Body cavity is a true coelom, often divided by internal septa.
- Body possesses a through gut with mouth and anus.
- Body possesses three separate sections: head, body, and pygidium.
- Body has a nervous system with an anterior nerve ring, ganglia and a ventral nerve cord.
- Body has a true closed circulatory system.
- Body has no true respiratory organs.
- Reproduction: normally sexual and gonochoristic or hermaphoditic.
- Trochophore larva is present during indirect development.

Phylum Annelida includes segmented worms. These animals are found in marine, terrestrial, and freshwater habitats. But the presence of water or humidity is a critical factor for their survival, especially in terrestrial habitats. The name of the phylum is derived from the Latin word annellus, which means a small ring. Animals in this phylum show parasitic and commensal symbioses with other species in their habitat. Approximately 16,500 species have been described in phylum Annelida. The phylum includes earthworms, polychaete worms, and leeches. Annelids show protostomic development in embryonic stages and are often called "segmented worms" due to their key characteristic of metamerism, or true segmentation.

Annelids display bilateral symmetry and are worm-like in overall morphology. Annelids have a segmented body plan wherein the internal and external morphological features are repeated in each body segment. Metamerism allows animals to become bigger by adding "compartments" while making their movement more efficient. This metamerism is thought to arise from identical teloblast cells in the embryonic stage, which give rise to identical mesodermal structures. The overall body can be divided into head, body, and pygidium (or tail). The clitellum is a reproductive structure that generates mucus that aids in sperm transfer and gives rise to a cocoon within which fertilization occurs; it appears as a fused band in the anterior third of the animal.

The epidermis is protected by an acellular, external cuticle, which, however, is much thinner than the cuticle found in the ecdysozoans and does not require periodic shedding for growth. Circular as well as longitudinal muscles are located interior to the epidermis. Chitinous hairlike extensions anchored in thc cpidcrmis and projecting from the cuticle, called setae/chaetae are present in every segment. Annelids show the presence of a true coelom derived from embryonic mesoderm and protostomy. Hence, they are the most advanced worms. A well-developed and complete digestive system is present in earthworms (oligochaetes) with a mouth, muscular pharynx, esophagus, crop, and gizzard being present. The gizzard leads to the intestine which ends in an anal opening. Each segment is limited by a membranous septum that divides the coelomic cavity into a series of compartments.

Annelids possess a closed circulatory system of dorsal and ventral blood vessels that run parallel to the alimentary canal as well as capillaries that service individual tissues. In addition, these vessels are connected by transverse loops in every segment. These animals lack a well-developed respiratory system, and gas exchange occurs across the moist body surface. Excretion is facilitated by a pair of metanephridia (a type of primitive "kidney" that consists of a convoluted tubule and an open, ciliated funnel) that is present in every segment towards the ventral side. Annelids show well-developed nervous systems with a nerve ring of fused ganglia present around the pharynx. The nerve cord is ventral in position and bears enlarged nodes or ganglia in each segment.

Annelids may be either monoecious with permanent gonads (as in earthworms and leeches) or dioecious with temporary or seasonal gonads that develop (as in polychaetes). However, cross-fertilization is preferred in hermaphroditic animals. These animals may also show simultaneous hermaphroditism and participate in simultaneous sperm exchange when they are aligned for copulation.

In conclusion, phylum Annelida includes vermiform, segmented animals. Segmentation is seen in internal anatomy as well, which is called metamerism. Annelids are protostomes. These animals have well-developed neuronal and digestive systems. Some species bear a specialized band of segments known as a clitellum. Annelids show the presence numerous chitinous projections termed chaetae, and polychaetes possess parapodia. Suckers are seen in class Hirudinea. Reproductive strategies include sexual dimorphism, hermaphroditism, and serial hermaphroditism. Internal segmentation is absent in class Hirudinea.

7-2 Class 1: Polychaeta

The vast majority of the more than 8,000 known species of polychaete worms are marine; some, however, are found in fresh or brackish water. They are abundant from the intertidal zone to depths of over 5,000 m. The polychaetes, so named because of the numerous setae (chaetae) they bear, range in length from less than 0.3 cm to more than 2.7 m, but most are from 5 cm to 10 cm long. Their colors are often brilliant, and some species are iridescent. The class has usually been divided on the basis of mode of existence into two groups, the Errantia and the Sedentaria.

Body is bilaterally symmetry with true coeloms (Fig. 7-1). Digestive system is complete and circulatory system is closed type. They have well-developed nervous system while excretory system consists of both metanephridia and protonephridia. The body does not have clitellum. The sexes are separate and fertilization is external. Development shows larval stage; larva is free-swimming, known as trochophore. Gonads are present in most of the body segments but some species do not contain gonads. Examples: *Neanthes succinea* (the sandworm), *Chaetopterus variopedatus* (the paddle worm) etc.

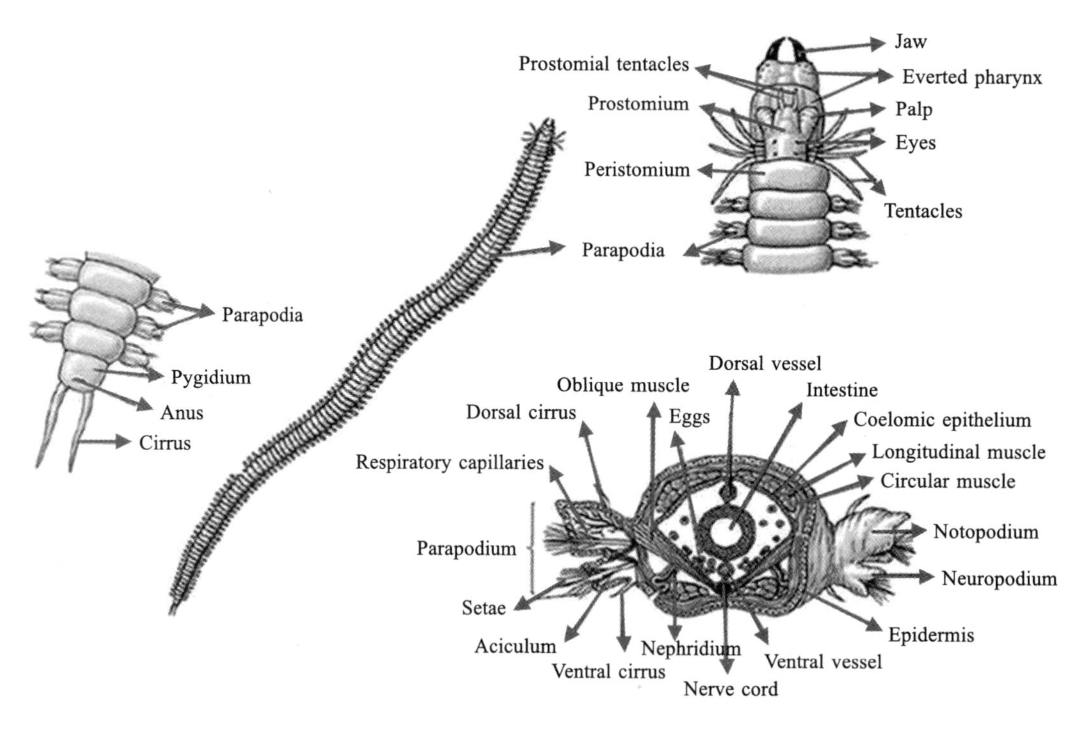

Fig. 7-1 Schematic of polychaetes

7-3 Class 2: Oligochaeta

The subclass Oligochaeta contains all the animals commonly thought of as "earthworms". Long thin worm has no obvious appendages to their bodies and greatly reduced heads so that when the animal is still, it is not sometimes obvious which end is the head and which is the tail. As a group, they are all morphologically similar and taxonomic division is often made far more on the basis of internal characteristics, particularly the positioning of the genitalia.

The Oligochaeta are the second largest group of the Annelida. With 3,100 known species, they make up about one third of the phylum. Within this diversity of species, there are aquatic forms, both freshwater and marine and also many terrestrial species. In terms of diet, the smaller species are often predatory while the larger species are soil or mud feeders. There are also a few parasitic species.

● **Nervous System**

The nervous system consists of a brain, which is found in the prostomium and connected to the ventral nerve cord (Fig. 7-2), and the sub-pharyngeal ganglia, by a special set of nerves called the circum-pharyngeal connectives. The ventral nerve cord is surrounded by a fibrous sheath. The nerve cord contains two sorts of nerve fibres: normal nerves and giant nerves. The giant nerves are only important during rapid escape manoeuvres when the animal needs to react very quickly. The ventral nerve cord runs the whole length of the animal and gives rise to several sets of nerves in each segment. In the more active forms, each segment has its own small ganglia, but in the more sedentary forms these are absent. Also, the more active the lifestyle of an animal, the larger is its brain, because of the need to interpret a much greater input of sensory information as well as the need to coordinate more varied and complicated movements.

● **Blood and Circulation**

The circulatory system is a closed system, meaning that they have blood vessels through which the blood flows. There are two main longitudinal blood vessels, a dorsal one and a ventral one, as well as three smaller longitudinal vessels, two lateral neural vessels and one sub-neural vessel. These vessels have circular muscles around them which can contract rhythmically to keep the blood moving around the body. The vessels also contain valves which ensure the blood only flows in one direction. Blood flows from the head to the tail in the ventral vessel and back, from the tail to the head, in the dorsal vessel. In each segment a number of smaller lateral (side) vessels branch off from the main vessels to supply the sections of that segment. The blood of annelids contains haemoglobin, the same respiratory pigment as in humans. It is this that makes their blood the same red colour as human's.

● **Gaseous Exchange**

Gaseous exchange normally occurs over the whole of the animal's body. Most aquatic oligochaetes, being smaller than polychaetes, have no need of special additional respiratory organs. Because gaseous exchange is far greater in the presence of moisture, terrestrial species secrete moisture in the form of coelomic fluid from the dorsal pores, mucous from the epidermal mucous glands and the excretions of the nephridia.

● **Excretion and Osmoregulation**

Excretion of metabolic wastes is through the action of nephridia. These are long coiled tubes which have many cilia lining their internal surface. In *Lumbricus*, these are longer and more complicated than those in the polychaetes and the tube is divided into three distinct parts, the innermost section of which is far wider than normal nephridial tubes. Both blood and coelomic fluid enter these nephridia where nutrients, water and salts are removed before the remaining wastes are passed out through the nephridiopore. Most body segments, except the first three and the last one, have their own pair of nephridia. Nephridia also serve as organs of osmoregulation. In species that live in fresh water or environments with variable salinity, the nephridial tubes are longer to help animals deal with the greater osmotic potential occurring between their inner body fluids and the fluids of the environment they are living in.

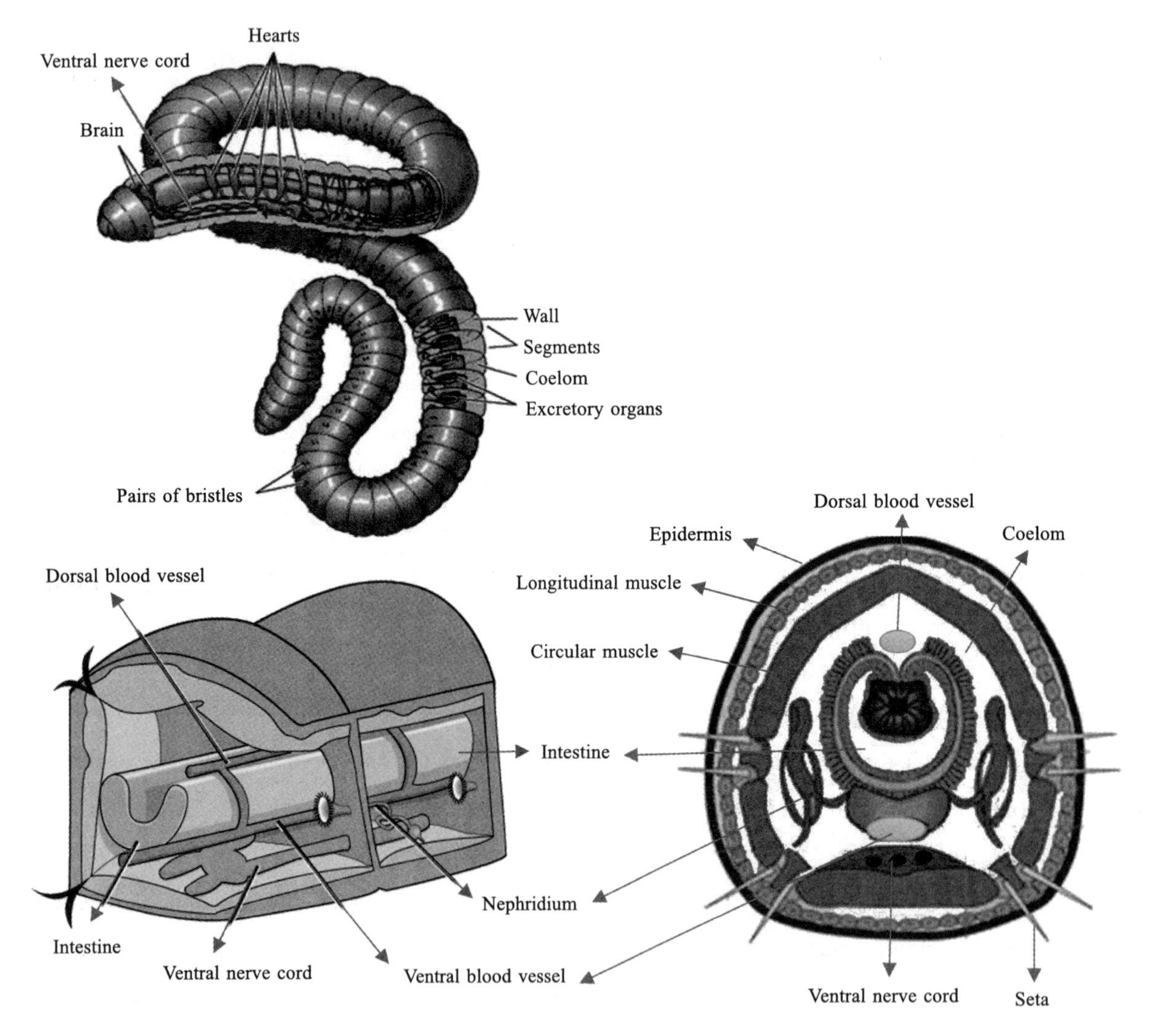

Fig. 7-2 Schematic of oligochaetes

- **Reproduction**

Oligochaetes are hermaphrodites, with separate testes and ovaries. The sexual organs, and the ducts that lead to and from them are situated in the anterior (front) part of the animal, normally between segments 7 and 15. The actual placement of the reproductive organs, including the openings of the ducts, which are normally on the same segments are important in classification. In *Lumbricus terrestris*, a common worm in Western Europe, the testes are in segments 10 and 11, the seminal vesicles in segments 9, 11, and 12, while the vas deferens opens on segment 15. The ovaries are found in segments 13 and the oviduct opens on segment 14. The sexual organs and their ducts are paired, one on each side of the worms' body. Sperm travels from the opening of the vas deferens to the clitellum, segments 32–36, along two seminal grooves (Fig. 7-3).

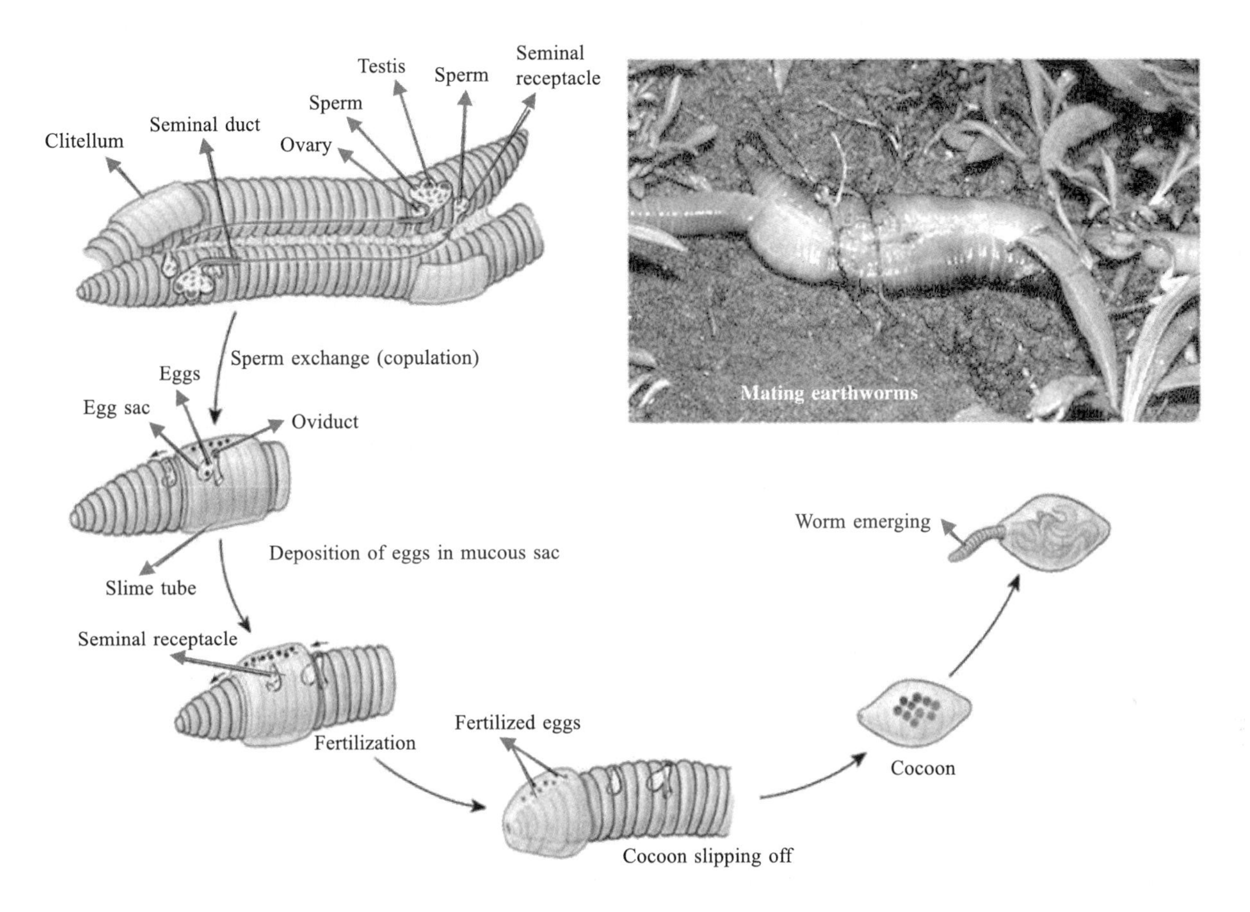

Fig. 7-3 Reproduction of earthworms

7–4 Class 3: Hirudinea

This class includes the 500 species of leeches, flattened, predacious or parasitic annelids equipped with suckers used for creeping. Leeches range in length from 1 cm to 20 cm; most are under 5 cm long. They are commonly black, brown, green, or red, and may have stripes or spots. Leeches are primarily freshwater annelids, but some live in the ocean and some in moist soil or vegetation. The majority of leeches are predators on small invertebrates; most swallow their prey whole, but some suck the soft parts from their victims. Some leeches are parasites rather than predators, and suck the body fluids of their victims without killing them. The distinction is not sharp, as many predatory leeches take blood meals on occasion.

● **Anatomy**

Leeches are the only annelids with a fixed number (34) of body segments; each segment has secondary subdivisions known as annuli. A clitellum, less conspicuous than that of oligochaetes, is present; there are no parapodia. A leech has a small anterior sucker and a larger posterior one (Fig. 7-4); the leech crawls by moving the anterior sucker forward, attaching it, and drawing up the posterior sucker. Most leeches can swim by rapid undulations of the body, using well-developed muscles of the body wall.

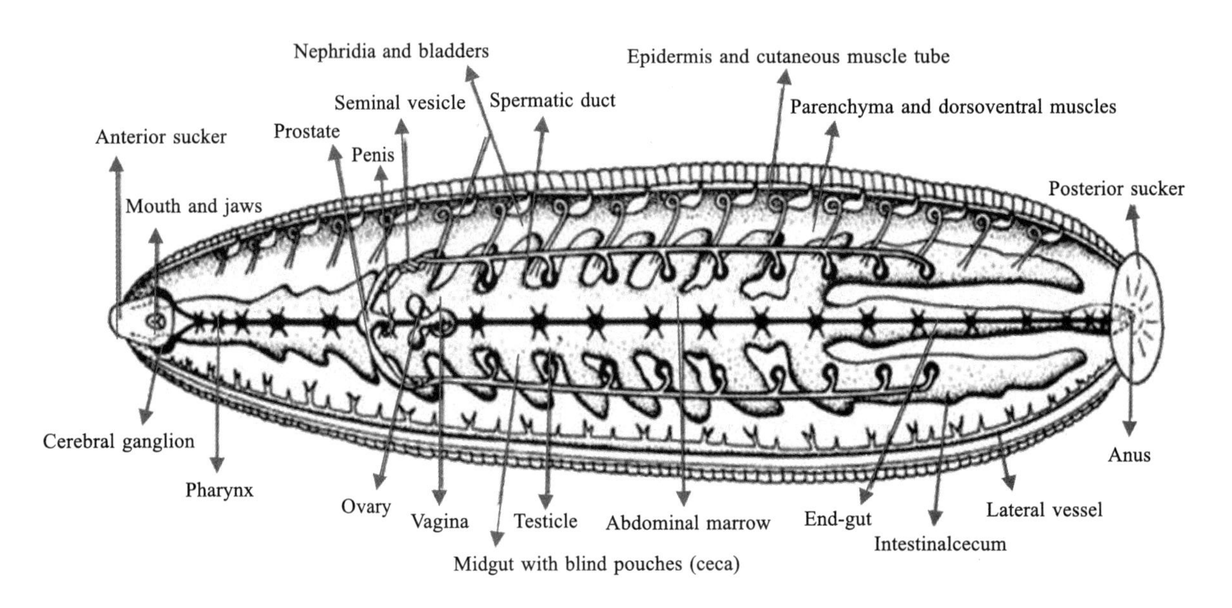

Fig. 7-4 Schematic of leeches

The coelom differs from that of other annelids in that it is largely filled in with tissue. Coelomic fluid is contained in a system of sinuses, which in some leeches functions as a circulatory system; there is a tendency in this group toward the loss of true blood vessels. The blood of some leeches is red. In others the blood lacks oxygen-carrying pigments and is therefore colorless; the oxygen dissolved directly in the blood is sufficient for respiration. Gas exchange occurs through the body surface of most leeches, although many fish-parasitizing leeches have gills.

The sense organs consist of sensory cells of various types, including photoreceptor cells, scattered over the body surface. There are also from 2–10 eyes, consisting of clusters of photoreceptor cells and located toward the front of the body.

● **Predation and Digestion**

Many leeches have a proboscis used for swallowing the prey or for sucking its fluids; others have jaws for biting. Many parasitic leeches are able to parasitize a wide variety of hosts. Most of the marine and some of the freshwater leeches are fish parasites. The medicinal leech, *Hirudo medicinalis*, is one of a group of aquatic bloodsucking leeches with jaws. Another group of jawed bloodsuckers is terrestrial; these leeches live in damp tropical vegetation and drop onto their mammalian prey. Most parasitic leeches attach to the host only while feeding; a single meal may be 5 or 10 times the weight of the leech and provides it with food for several months. The digestive tract of bloodsuckers produces an anticoagulant, hirudin, which keeps the engorged blood from clotting. A few leeches attach permanently to the host, leaving only to reproduce. Predatory leeches are active at night and hide by day.

● **Reproduction**

Like the oligochaetes, leeches are hermaphroditic and cross-fertilizing, although fertilization is internal. In some species, sperms are enclosed in sacs, called spermatophores, that are attached to the outside of the partner; sperms pass through the body wall to the ovaries, where the eggs are fertilized. In other species, sperms are not enclosed and are transferred directly into the body of the partner by copulation. A courtship display is seen among some leeches at the time of mating. The fertilized eggs are deposited in a cocoon, secreted by the clitellum; the cocoon is buried in mud or affixed to submerged objects. The young emerge as small copies of the adults.

7-5 学习要点

● 同律分节和次生体腔出现的生物学意义；后肾管、闭管式循环系统、链状神经系统、刚毛或疣足等器官功能和结构的一致关系。

● 无脊椎动物后肾管的三种起源；担轮幼虫trochophore的结构、发育和变态。

● 代表动物——环毛蚓*Pheretima*的形态结构特征、生殖和个体发育、经济意义。

● 代表动物——沙蚕*Nereis*的外部形态结构特征（口前叶、口前触手、围口触手、疣足的基本构造等）；多毛纲上述结构的主要功能。沙蚕的生殖习性（异沙蚕体heteronereis phase、群浮和婚游）；生殖习性与环境条件的适应关系。多毛类的经济意义及其与水产养殖和渔业的关系。

● 代表动物——医蛭*Hirudo*的生活方式及与之相适应的基本结构；血窦的来源及其与次生体腔的关系。

● 次生体腔的发生及其在动物演化史中的意义，以及其对动物身体机能所产生的深刻影响。

● 同律分节。

● 后肾管与原肾管的区别

● 沙蚕形态结构与蚯蚓的区别。

● 疣足的形态和结构。

7-6 巩固测验

【名词】

同律分节、异律分节、次生体腔（真体腔）、闭管式循环、刚毛、疣足、后肾管、索式神经系统（链状神经系统）、担轮幼虫、项器、血窦

【选择】

1. 以下不属于环节动物门的是（　　）。

A. 多毛纲　　B. 寡毛纲　　C. 蛭纲　　D. 绦虫纲

2. 以下具担轮幼虫期的动物是（　　）。

A. 沙蚕　　B. 水蚓属　　C. 水蛭　　D. 涡虫

3. 以下不属于环节动物的运动器官是（　　）。

A. 疣足　　B. 刚毛　　C. 吸盘　　D. 鞭毛

4. 环节动物的生殖细胞来源于（　　）。

A. 内胚层　　B. 外胚层　　C. 体腔膜　　D. 内胚层的间细胞

5. 蚯蚓的受精囊是（　　）。

A. 卵子受精的场所　　B. 贮存自体卵子的场所

C. 贮存自体精子的场所　　D. 贮存异体精子的场所

6. 从动物界演化看，从（　　）开始，消化管中除肠上皮外还具有肌肉层，能同时进行物理性消化。

A. 扁形动物　　B. 软体动物　　C. 环节动物　　D. 线形动物

7. 环节动物比线形动物高等的特征是（　　）。

A. 异律分节、次生体腔、后肾管　　B. 同律分节、初生体腔、原肾管

C. 同律分节、次生体腔、后肾管　　D. 异律分节、次生体腔、原肾管

8. 在进化过程中，真体腔首先出现于（　　）。

A. 腔肠动物　　B. 线形动物　　C. 环节动物　　D. 节肢动物

9. 下列生物中，身体不分节的是（　　）。

A. 蚯蚓　　B. 蝗虫　　C. 蛔虫　　D. 海盘车

10. 蚯蚓的运动器官是（　　）。

A. 纤毛　　B. 鞭毛　　C. 刚毛　　D. 疣足

11. 蚯蚓的排泄器官是（　　）。

A. 原肾管　　B. 后肾管　　C. 马氏管　　D. 收集管

12. 蚯蚓暴露于干燥空气中，见到阳光，很快就会死亡，其原因是（　　）。

A. 体内失水过多

B. 体内废物不能及时排出

C. 体表失水、干燥，影响呼吸而窒息死亡

D. 蚯蚓是厌氧型的，接触到氧气后无氧呼吸受到抑制而死

13. 在蚯蚓的循环系统中，血液流动的方向是（　　）。

A. 背血管→心脏→腹血管　　B. 腹血管→心脏→背血管

C. 心脏→背血管→腹血管　　D. 背血管→腹血管→心脏

14. 在蠕虫中，肠壁上具有平滑肌的是（　　）。

A. 马陆　　B. 蚂蟥　　C. 线虫　　D. 涡虫

15. 蚯蚓的排泄器官是（　　）。

A. 后肾　　B. 原肾　　C. 后肾管　　D. 原肾管

【填空】

1. 蚯蚓的中枢神经包括（　　　）、（　　　）、（　　　）、（　　　），为典型的（　　　）神经系统。

2. 蚯蚓的体壁包括（　　　）、（　　　）、（　　　）、（　　　）、（　　　）五层。

【简答】

1. 环节动物门的主要特征有哪些？蚯蚓适应土壤穴居生活的主要外部特征有哪些？

2. 分节的意义是什么？真体腔出现的意义是什么？

3. 后肾管与原肾管有什么区别？

4. 真体腔与假体腔有什么区别？

5. 简述环毛蚓的循环系统特点。

Chapter 8 Phylum Mollusca

8-1 Characteristics of Phylum Mollusca

- They have a bilaterally symmetrical body (exception: gastropods have a asymmetrical body) and show organ system grade of organization.
- They are mostly aquatic organisms. Most of them inhabit in the marine environment and few are freshwater. Besides these, some live in terrestrial damp soil.
- They are triploblastic animals with unsegmented (exception: *Neopilina*) soft body.
- Mantle and shell cover the body and the body has three regions namely head, ventral foot, and a visceral mass. In this case, the shell is a hard calcareous structure, made up of calcium carbonate.
- They have open type blood circulatory system with heart (one or two auricles and one ventricle) and aorta. The body cavity is known as hemocoel which circulates blood.
- They have a well-developed and complex digestive system. The mouth contains a rasping structure "radula" with chitinous teeth for scraping or cutting food.
- The nervous system consists of three pairs of ganglia (cerebral, visceral and pedal) with connectives and nerves.
- In aquatic mollusks, respiration occurs through gills or ctenidia which are located in the mantle cavity while in terrestrial forms, it occurs through lungs (pulmonary sacs) or general body surface.
- They have metanephridia (kidney) for excretion.
- Sense organs consist of eyes, osphradium, tentacles, and statocyst.
- Sexes are usually separate but some are monoecious. In this case, sexual reproduction occurs.
- External or internal fertilization occurs.
- They show direct or indirect development with trochophore or veliger larval stages. Glochidium larva also occurs during the parasitic stage for some species (mussels and clams).
- They use the ventral muscular foot for locomotion.
- Phylum Mollusca can be divided into the following different classes: Pelecypoda or Bivalvia, Gastropoda, Cephalopoda, Aplacophora, Monoplacophora, Polyplacophora, Scaphopoda.

Mollusca makes the second largest phylum of non-chordate animals including snails, octopuses, sea slugs, squid, and bivalves such as clams, oysters, and mussels. This phylum contains about 100,000 described species. Among all known marine species, 23% are mollusks. But some species live in freshwater and terrestrial habitats. This group displays a broad range of morphological features such as a muscular foot, a mantle and visceral mass containing internal organs. Calcium carbonate secretes from the mantle and forms outer calciferous body shell in most of the mollusks. The size of the molluscans varies from less than 1 mm to 20 m. They play an important role in the life of humans because they are the source of food for many people as well as jewellery. Many molluscans are not good for human's life. Some cause diseases or acts as pests like the snails and slugs. Generally, the hard calciferous shells of mollusks are used to build awesome jewelry pieces. Some mollusks such as bivalves and gastropods produce valuable pearls. Natural pearls are produced when a small foreign object gets trapped in between the mollusk's body shell and mantle. Besides these, many scientists use bivalve mollusks as bioindicators of the freshwater and marine environments.

8-2 Class 1: Pelecypoda or Bivalvia

This class comprises more than 15,000 species including mussels, oysters, clams, scallops, etc. The largest living bivalve species is giant clam (*Tridacna gigas*) which can grow up to 1,200 mm in length and more than 200 kg in weight. No system of classification erected for the Bivalvia has been accepted by all. Paleontologists interpret bivalves on the basis of shell features, notably shell and ligament structure, arrangement of hinge teeth, and body form as interpreted from internal muscle scars.

● **Shell**

The bivalve shell is made of calcium carbonate embedded in an organic matrix secreted by the mantle. The periostracum, the outermost organic layer, is secreted by the inner surface of the outer mantle fold at the mantle margin. It is a substrate upon which calcium carbonate can be deposited by the outer surface of the outer mantle fold. The number of calcareous layers in the shell (in addition to the periostracum), the composition of those layers (aragonite or aragonite and calcite), and the arrangement of these deposits (e.g. in sheets, or foliate) is characteristic for different groups of bivalves. Middorsally an elastic ligament creates the opening thrust that operates against the closing action of the adductor muscles (Fig. 8-1). The ligament typically develops either externally (parivincular) or internally (alivincular) but comprises outer lamellar and inner fibrous layers secreted by the mantle crest. The ligament type is generally characteristic of each bivalve group. The hinge plate with ligament also possesses interlocking teeth to enforce valve alignment and locking, when closed, to prevent shear. Many variations in teeth structure occur.

Pearls are formed when an irritant, such as a bit of food, a grain of sand, bacteria, or even a piece

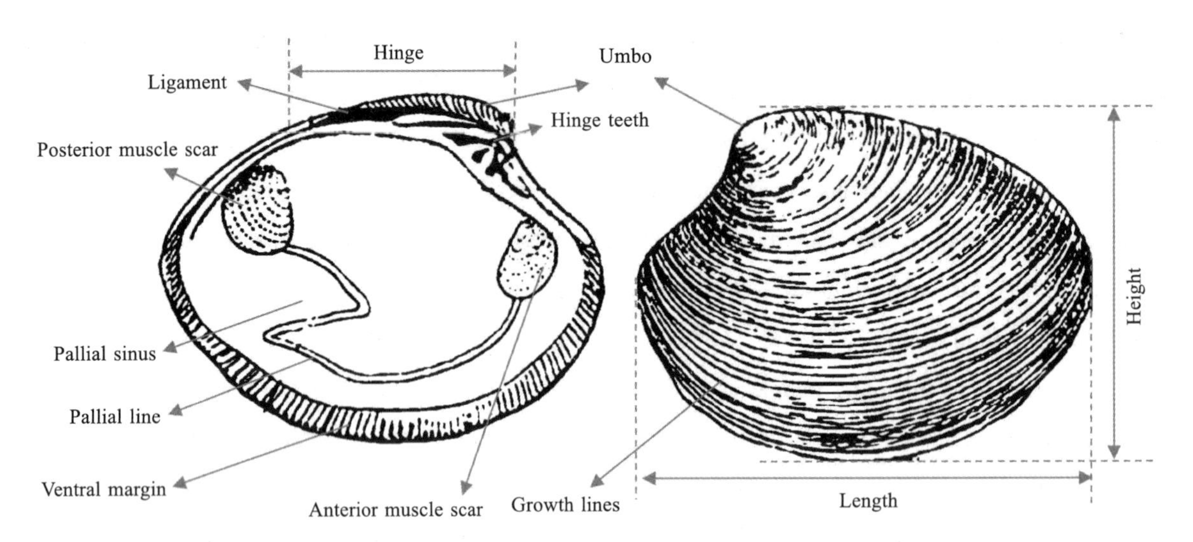

Fig. 8-1 Shell of bivalves

of the mollusk's mantle becomes trapped in the mollusk. To protect itself, the mollusk secretes the substances aragonite and conchiolin (a protein), which are the same substances it secretes to form its shell. The composite of these two substances is called nacre, or mother-of-pearl. The layers are deposited around the irritant and it grows over time, forming the pearl (Fig. 8-2).

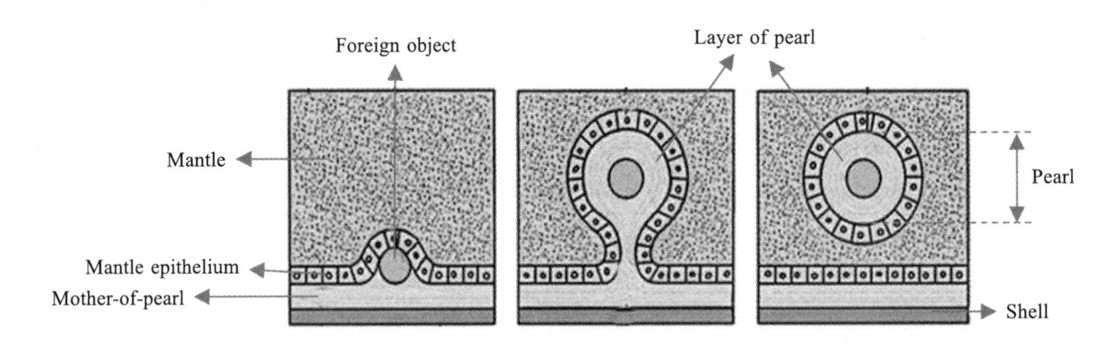

Fig. 8-2 Formation of the pearl

● **Mantle and Musculature**

The mantle lobes secrete the shell valves; the mantle crest secretes the ligament and hinge teeth. Growth takes place at the margins, although increases in thickness take place everywhere. The mantle is withdrawn between the shell valves by mantle retractor muscles, their point of attachment to the shell being called the pallial line. The musculature comprises two (dimyarian) primitively equal (isomyarian) adductor muscles: the anterior and the posterior. The anterior of these may be reduced (anisomyarian; heteromyarian) or lost (monomyarian). Only very rarely is the posterior lost and the anterior retained.

● **Nervous System and Organs of Sensation**

The nervous system is simple and the head is completely absent, reflecting the sedentary habit. In primitive bivalves (e.g. Palaeotaxodonta), there are four pairs of ganglia—cerebral, pleural, pedal, and visceral. In all other bivalves, the cerebral and pleural ganglia are fused into two cerebropleural ganglia, located above and on either side of the esophagus. The pedal ganglia are in the base of the foot, and the visceral ganglia are located under the posterior adductor muscle. Nerve fibres arising from the cerebropleural ganglia extend to the pedal and visceral ganglia. In some bivalves with long siphons, there are accessory siphonal ganglia. In many swimming bivalves the visceral ganglia are much enlarged, presumably to coordinate complex swimming actions.

● **Digestive System and Nutrition**

The bivalve digestive system comprises a complex stomach and associated structures but an otherwise simple intestine (Fig. 8-3). The various types of stomach have been used to erect an alternative classification. Digestion typically takes place in two phases: extracellular in the stomach and intracellular in the digestive diverticula, opening laterally from the stomach wall. Transport of food particles is affected by cilia, creating an array of tracts and sorting areas within the stomach. The principal organ of extracellular digestion is the crystalline style. It is rotated in its sac by cilia; the head, projecting into the stomach, grinds against a part of the stomach wall lined by a chitinous gastric shield. As it rotates, it dissolves, releasing enzymes and initiating primary extracellular digestion of the mucus-bound food. Products of this process are passed in a fluid suspension into large embayments and thence into the digestive diverticula, where intracellular digestion takes place. Waste material is consolidated in the midgut and rectum and expelled as firm fecal pellets from an anus opening into the exhalant stream. Feeding and digestion are highly coordinated, typically regulated by tidal and diurnal cycles.

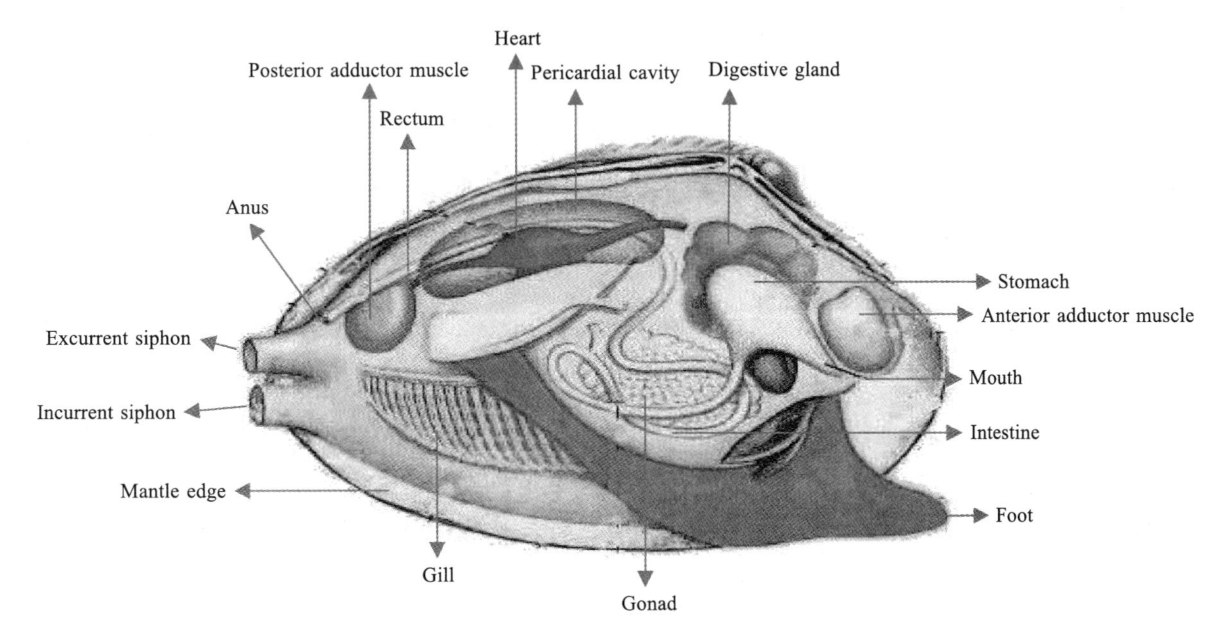

Fig. 8-3 Schematic of bivalve

● **Excretory System**

Blood is forced through the walls of the heart into the pericardium. From there it passes into the kidneys where wastes are removed, producing urine. The paired kidneys (nephridia) are looped with an opening into the pericardium and another into the suprabranchial chamber. The kidneys may be united. Bivalves also possess pericardial glands lining either the auricles of the heart or the pericardium; they serve as an additional ultrafiltration device.

● **Respiratory System**

In the primitive bivalves, the paired gills are small and located posteriorly. The gills in all other bivalves (save septibranchs, which have lost their gills) are greatly enlarged and possess a huge surface area. While the gills are thought to serve a respiratory function, respiratory demands are low in these mostly inactive animals, and, since the body and mantle are both bathed in water, respiration probably takes place across these surfaces as well. Such a mechanism has been demonstrated for a few bivalves, most notably freshwater species that are exposed to occasional drought. In such species, drying induces slight shell gaping posteriorly, the mantle margins exposing themselves to air. For most intertidal bivalves (which are alternately exposed to wetting and drying), respiration all but ceases during the drying phase.

● **Vascular System**

The heart, enclosed in a pericardium, comprises a medial ventricle with left and right auricles arising from it. Blood oxygenated within the ctenidia flows to the auricles and from there to the ventricle, where it is pumped into anterior and posterior aortas. The blood then enters hemocoelic spaces in the mantle and visceral mass and returns to the heart via the ctenidia or the kidneys. The blood serves both to transport oxygen and metabolic products to tissues deep within the body and as a hydrostatic skeleton (for example, in the extension of the foot during locomotion and siphons during feeding). There are amoeboid corpuscles, but, except in a few bivalves, no hemoglobin or other respiratory pigment occurs.

● **Reproductive System**

The reproductive system is simple and comprises paired gonads. These gonads discharge into the renal duct in primitive bivalves but open by separate gonopores into the suprabranchial chamber in more modern bivalves. Typically, the sexes are separate, but various grades of hermaphroditism are not uncommon. Eggs and sperms are shed into the sea for external fertilization in most bivalves, but inhalation of sperms by a female permits a type of internal fertilization and brooding of young, usually within the ctenidia.

8–3 Class 2: Gastropoda

Gastropods, more than 65,000 animal species belonging to the class Gastropoda, form the largest group in the phylum Mollusca. The class is made up of the snails, which have a shell into which the animal can generally withdraw, and the slugs, which are snails whose shells have been reduced to an internal fragment or completely lost in the course of evolution (Fig. 8-4).

Gastropods are among the few groups of animals to have become successful in all three major habitats: the ocean, fresh waters, and land. A few gastropod types (such as conches, abalones, limpets, and whelks) are used as food, and several different species may be used in the preparation of escargot. Very few gastropod species transmit animal diseases; however, the flukes that cause human schistosomiasis use gastropods as intermediate hosts. The shells of some species are used as ornaments or in making jewelry. Some gastropods are scavengers, feeding on dead plant or animal matter; others are predators; some are herbivores, feeding on algae or plant materials; and a few species are external or internal parasites of other invertebrates.

● **Body**

The gastropod body consists of four main parts: visceral hump, mantle, head, and foot. The body is attached to the shell (Fig. 8-5) either by one columellar muscle or by a series of muscles. Typical snails can withdraw the head and foot into the shell, but numerous species have shells so reduced in size as to be unable to contain the body; slugs, of course, have either an internal shell vestige or no shell at all.

● **Visceral Hump**

The visceral hump, or visceral mass, of gastropods is always contained within the shell (Fig. 8-6); it generally holds the bulk of the digestive, reproductive, excretory, and respiratory systems. A significant part of the visceral hump consists of the mantle, or pallial, cavity. In both prosobranchs and shelled opisthobranchs this is a cavity completely open anteriorly; in pulmonates it is closed except for a narrow pore. The mantle tissue at the forward edge of the cavity secretes the shell. The upper surface of the mantle cavity serves a respiratory function. In marine species the ciliated lining of the mantle cavity helps produce a water current that passes posteriorly across the gill, or ctenidium, and the osphradium, which is thought to be a sensory receptor that can detect chemical changes in the environment. Both organs lie on the left anterior side of the cavity. The water current sweeps across the posterior part of the mantle cavity, where the nephridiopore, or kidney opening, lies; the water current then passes anteriorly along the right margin past the anus, through which undigested particles of food are eliminated, and usually moves past the gonopore, through which sexual products are released. Cilia on the gill play an important role in water flow through the mantle cavity; they also help some species

Fig. 8-4 Gastropods

(e.g. *Crepidula*) capture food particles.

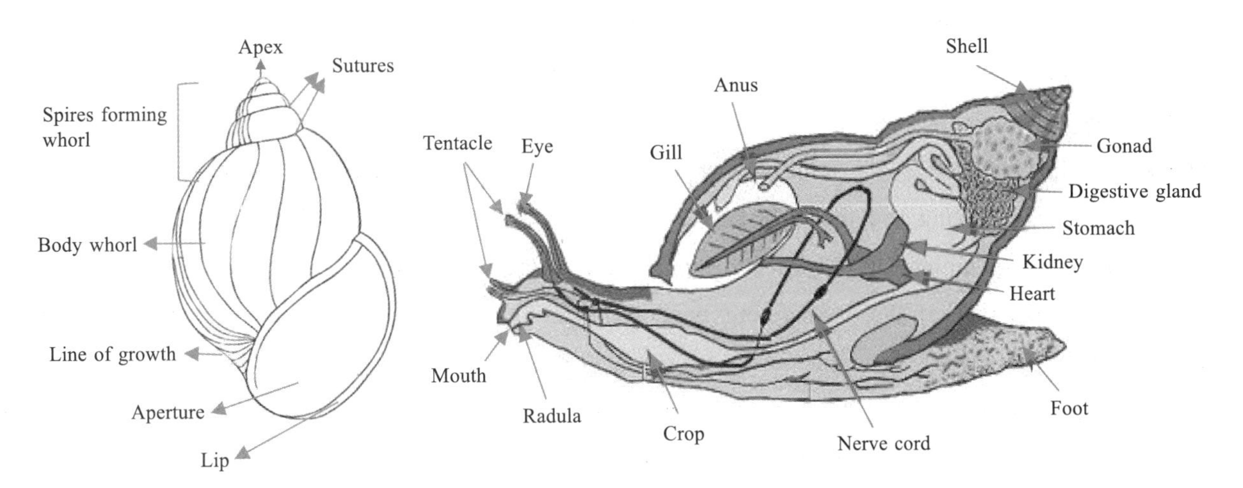

Fig. 8-5 Shell of gastropods

Fig. 8-6 Schematic of gastropods

The mantle cavity serves as a space for the head and foot when these organs are retracted. Many land pulmonates apparently also use the mantle cavity to retain water. Prosobranchs use the operculum, the horny or calcareous disk located on the back of the foot at the posterior end, to seal the shell opening after the head and body have been retracted.

● **Nervous System and Sense Organs**

A series of paired ganglia (knotlike masses of nerve cell bodies that collectively function as the central nervous system) are connected by nerve cords, which are bilaterally arranged in the primitive forms. The process of torsion has twisted the visceral cords into the form of a figure eight. In more-advanced gastropods there are secondary modifications to a more nearly bilateral state, and in many groups they have been detorsion. Water-dwelling mollusks depend primarily upon ciliary water currents passing across chemoreceptors for information from the environment. The primary chemoreceptors in the gastropod body are scattered over the skin surface, protruding from tentacles or palps, and housed inside the mantle cavity in the form of the osphradium, an olfactory organ connected to the respiratory system. Sense organs are more highly developed in carnivores than in herbivores. Eyespots, located at the base (most gastropods) or tip (land pulmonates) of the eye tentacles, are primarily light-sensitive rather than image-forming. A pair of statocysts, thought to be balancing organs, is present in nonsessile taxa.

● **Digestive System**

The radular motion conveys food particles into the mouth, and ciliary water currents move the food through the digestive tract, except in carnivores, where muscular action plays an important role. Various salivary and digestive glands secrete enzymes into either or both the buccal cavity and the stomach, where digestion takes place. The apical digestive gland, or "liver", can store digested food for use during periods of inactivity.

● **Excretory System**

There are two kidneys, or nephridia, in primitive gastropods, such as the archaeogastropods, while, in the advanced forms, one kidney is small or lost. The kidney plays different roles, depending upon the environment in which the snail lives. Most marine gastropods have the same total concentrations of solutes as in the surrounding seawater, and thus a small osmotic differential (i.e. an equilibrium) exists between the water leaving and that entering the cell. Little energy is needed therefore to prevent the cells from losing or gaining too much water. Freshwater gastropods, however, have a higher total solute concentration than that of the surrounding water. The kidney must expend energy to control water balance (osmoregulation). The flow of water through the mantle cavity is restricted in freshwater species by the closure of the mantle cavity by the mantle collar. Land prosobranchs have an open mantle cavity and, in order to conserve water, secrete nearly crystalline urine. Land pulmonates have a ureteric groove or closed ureter that resorbs water from the urine. In both marine and freshwater species, ciliary water currents sweep the excreted matter out of the mantle cavity.

● **Respiration**

In marine and freshwater gastropods, respiration takes place as water currents pass across the gill surfaces within the mantle cavity in most species with spiral shells, across gill elements along the sides of the bodies in most limpets, or through projections from the body surface in sea slugs or other taxa with reduced shells. The upper surface of the mantle cavity is heavily vascularized in land snails, which use muscular contractions to pump air in and out of the small respiratory pore at the anterior edge of the mantle cavity. In some land slugs or tropical snails with reduced shells, respiratory functions have shifted either to external projections from the mantle collar or to the skin as the area of mantle roof available for respiration has decreased in size.

● **Reproductive System and Life Cycles**

The primitive archaeogastropods retain two nephridia; the right nephridium provides the passage for eggs or sperm from the ovary or testis to the mantle cavity. The sexes are separate in nearly all prosobranchs, although in a few taxa, such as *Crepidula*, animals begin life as males and then change to females later. Opisthobranch and pulmonate species are hermaphroditic and often protandrous (male gonads maturing first); however, in many taxa, adults become simultaneous hermaphrodites (male and female gonads are functional at the same time). Internal fertilization is common in the more advanced marine species but mandatory in the freshwater and terrestrial groups. A very few gastropod species are parthenogenetic (gametes developing without fertilization); the progeny of these species are clones of the parent.

Gastropods originated in the oceans, and relics of this fact are preserved in the early life history of freshwater and land species. Only in the most primitive prosobranchs (such as abalone) are the gametes released into the water for fertilization to take place outside the female. The fertilized egg hatches into a free-swimming form (trochophore larva). Upon the expansion of the ciliary girdle of the trochophore larva into large, heavily ciliated lobes (vela), the larva, called a veliger, undergoes torsion, a 180° twisting of the body that brings the posterior part of the body to an anterior position behind the head. Torsion is unique to the gastropods.

8–4 Class 3: Cephalopoda

Cephalopods, are members of the class Cephalopoda of the phylum Mollusca, a small group of highly advanced and organized, exclusively marine animals. The octopus, squid, cuttlefish, and chambered nautilus are familiar representatives (Fig. 8-7). The extinct forms outnumber the living, the class having attained great diversity in late Paleozoic and Mesozoic times. The extinct cephalopods are the ammonites, belemnites, and nautiloids, except for five living species of *Nautilus*.

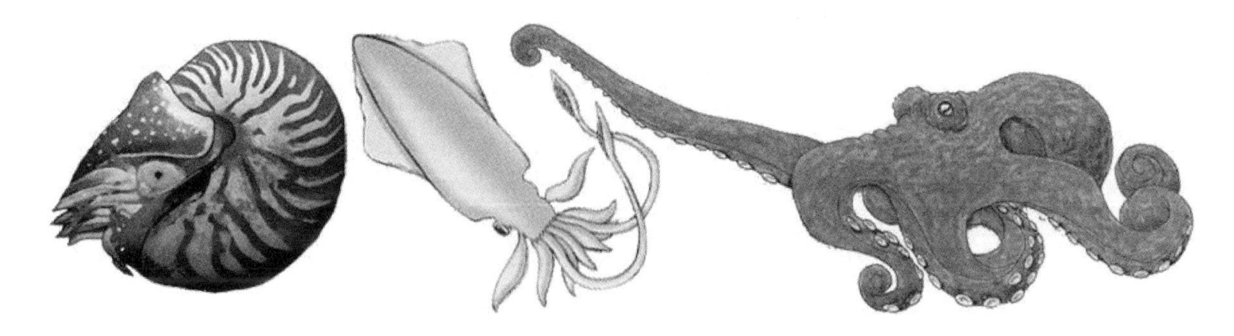

Fig. 8-7 Cephalopods

● **Organs**

All cephalopods have similarities in certain organs. They have three hearts, two of which move blood to the gills while the other pumps blood to the rest of the body. Their blood is blue, because it binds oxygen with the protein hemocyanin. Their brains are much larger than those of other invertebrates and most species are able to learn and remember information. They have two eyes, which are generally characterized as being extremely complex, perhaps even as sophisticated as the human eye (Fig. 8-8).

● **Color**

Cephalopods are able to change the color of their skin very rapidly and at will. They can even make intricate patterns and shapes on their skin. This is achieved using chromatophores, which are bags filled with pigment located in the skin. These chromatophores can be controlled by nerves to change the color of the skin. They primarily use this ability to help camouflage themselves, but it's also used as part of mating rituals.

● **Anatomy**

All cephalopods have the same basic anatomy. They are made up of a body, a head and a foot.

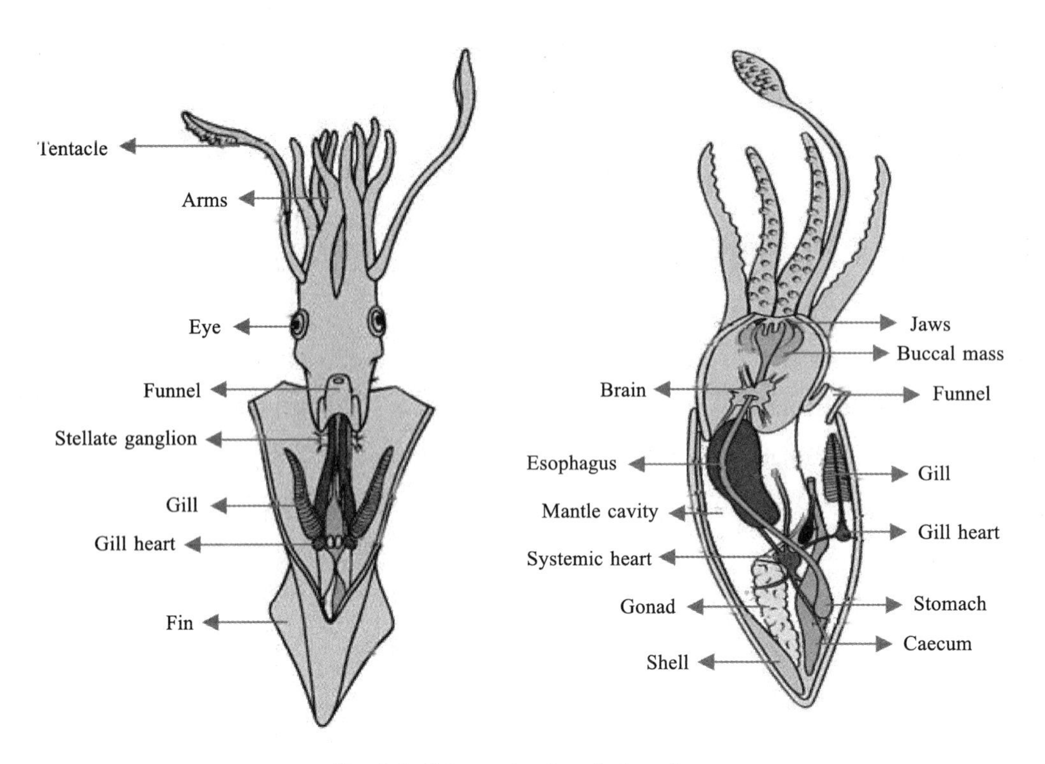

Fig. 8-8 Schematic of cephalopods

They have a muscular casing called a mantle which contains and protects their organs. They all have arms—at least eight of them—that are attached directly to their heads, but only some species also have tentacles. Their arms will either have cirri, suckers or hooks on their undersides.

● **Diet**

Cephalopods are all strict carnivores. Exactly what they eat depends on the species and its size, but common prey includes various fishes, crustaceans and mollusks. They have hard beaks, made out of horn, which are useful for tearing at and devouring their prey. While most species hunt for their food, there are certain species that prefer to scavenge for their meals.

8-5 学习要点

● 软体动物血窦 sinus 的来源。

● 间接发育（海水种类发育中一般须经担轮幼虫和面盘幼虫阶段）。

● 代表动物——河蚌 *Anodonta* 的形态结构特征、生活习性和瓣鳃的构造；珍珠形成的原理及过程。

● 贝壳的形态构造、方位和测量；腹足纲左右不对称的起源；腹足纲的重要类群（前鳃亚纲、后鳃亚纲和肺螺亚纲）；主要经济种类和习见种类如鲍 *Haliotis*、红螺 *Rapana*、笠贝、马蹄螺、田螺、海牛、蜗牛、椎实螺等的分类地位和经济意义。

● 主要的经济种类如三角帆蚌 *Hyriopsis cumingii*、牡蛎 *Ostrea*、缢蛏 *Sinonovacula constricta*、蚶 *Arca*、江珧 *Pinna*、扇贝 *Chlamys*、贻贝 *Mytilus*、珍珠贝 *Pteria* 等的分类地位和经济意义。

● 头足纲的特征（身体分为头部、足部和胴部，腕可分化为茎化腕和触腕，肉鳍在胴体上可分化为周鳍型、中鳍型和端鳍型）；头足类在外形上适应快速游泳的特点；柔鱼 *Ommatostrephes*、乌贼 *Sepia* 和枪乌贼 Loliginidae 的分类地位、经济意义和分布。

8-6 巩固测验

【名词】

外套膜、血窦、鲍雅氏器、凯伯尔氏器、钩介幼虫、晶杆、海螵蛸。

【选择】

1. 下列不属于外套膜功能的是（　　）。
A. 分泌形成贝壳　B. 形成外套腔　C. 形成出入水管　D. 感觉中心

2. 下列器官，与软体动物的消化无关的是（　　）。
A. 齿舌　B. 晶杆　C. 胃盾　D. 外套膜

3. 圆田螺的最终排泄产物为（　　）。
A. 尿素　B. 氨基酸　C. 氨　D. 尿酸

4. 乌贼的漏斗结构由（　　）特化而成。
A. 足　B. 外套膜　C. 内脏团　D. 触角

5. 软体动物中种类最多的一个类群是（　　）。
A. 腹足纲　B. 双壳纲　C. 头足纲　D. 多板纲

6. 软体动物有排泄功能的结构称为（　　）。
A. 心肾复合体　B. 马氏管　C. 原肾管　D. 围心腔

7. 无齿蚌的幼虫称为（　　），可以寄生在鱼的鳃和鳍等处。
A. 钩介幼虫　B. 牟勒氏幼虫　C. 担轮幼虫　D. 浮浪幼虫

8. 乌贼的排泄废物为（　　）。
A. 尿酸　B. 鸟嘌呤　C. 氨　D. 尿素

9. 圆田螺的生殖方式为（　　）。
A. 卵生　B. 卵胎生　C. 胎生　D. 孤雌生殖

10. 头足类的卵裂方式为（　　）。
A. 盘状卵裂　B. 螺旋卵裂　C. 辐射卵裂　D. 表面卵裂

11. 圆田螺的内部神经（　　）。
A. 发生扭转，呈“8”形　B. 发生反扭转，不呈“8”形
C. 不发生扭转，呈“8”形　D. 发生反扭转，呈“8”形

12. 下列动物被称为“活化石”的是（　　）。

A. 鹦鹉螺　　B. 龙女簪　　C. 虎斑宝贝　　D. 珍珠贝

【填空】

1. 判断螺壳左右旋的方法是将壳顶向（　　　　），壳口面向（　　　　），壳口在螺轴左侧的为（　　　　），壳口在螺轴右侧的为（　　　　）。

2. 贝壳的珍珠层是由（　　　　）分泌形成的。

【简答题】

1. 软体动物的基本特征有哪些?

2. 珍珠的是怎样形成的?

3. 软体动物分类及代表动物各有哪些?

4. 举例说明软体动物呼吸器官的类型。

5. 举例说明软体动物排泄器官的类型。

Chapter 9 Phylum Arthropoda

9-1 Characteristics of Phylum Arthropoda

- The body is triploblastic, segmented, and bilaterally symmetrical.
- They exhibit organ system level of organization.
- The body is divided into head, thorax, and abdomen.
- Their body has jointed appendages which help in locomotion.
- The coelomic cavity is filled with blood.
- They have an open circulatory system.
- The head bears a pair of compound eyes.
- The exoskeleton is made of chitin.
- The terrestrial arthropods excrete through Malpighian tubules while the aquatic ones excrete through green glands or coxal glands.
- They are unisexual and fertilization is either external or internal.
- They have a well-developed digestive system.
- They respire through the general body surface or trachea.
- They contain sensory organs like hairs, antennae, simple and compound eyes, auditory organs, and statocysts.
- The living Arthropoda can be divided into 3 subphyla, 16 classes and innumerable orders and families.

Among the living animals of the world, crabs, prawns, woodlice, spiders, scorpions, insects, millipedes and centipedes are all arthropods, linked together by the possession of a hard jointed exoskeleton, a through-gut and jointed limbs.

Arthropods are currently thought to have evolved from Annelids. Both groups have the same sort of central nervous system, a similar circulatory system along with metameric segmentation and tagmatization.

The phylum Arthropoda is huge in terms of both numbers of species and numbers of individuals.

They have diversified to live in every habitat imaginable, from the tropics to the poles, from the bottom of the oceans to the tops of mountains, both underground and inside other animals and plants. Wherever you look, arthropods are ubiquitous. Most of the invertebrates you encounter during your life will be arthropods. With an amazing 1 million named species (and estimates of total species numbers rising to 30 million), the arthropods represent over 80% of the animal kingdom and probably at least half of all living organisms.

Arthropods are amazingly diverse in form and function and in many cases fundamental characteristics have been secondarily lost, either completely or are only visible in embryonic form. Most of these problems are generated by the Crustacea whose variability is incredible.

Because of their huge numbers, and the density at which they occur in many habitats, on land, in the soil, in fresh water and in the sea, arthropods are of immense importance to the ecology of the whole planet. It is true to say that without them, complicated multicellular life on this planet would simply collapse and probably disappear all together. Their economic importance to mankind is also beyond measure. They are important in nutrient recycling in both aquatic and terrestrial environments, comprising the key workers in most ecologies. They supply food directly for huge numbers of amphibians, fish, birds, mammals and reptiles and indirectly for more still. As insects, their value as pollinators of flowering plants and therefore as preservers of floral diversity is incalculable, while their contribution to modern biological and ecological research is equally extensive.

Arthropods have been around for a long time and several major and many minor lineages are now extinct, perhaps the most well-known of which are the trilobites. The first arthropods evolved in the warm seas of the Cambrian period about 540 million years ago. Amongst the first arthropods were animals called euthycarcinoids, creatures that appear to have been halfway between insects and crustaceans. Trace fossils, fossils left by animal activities, indicate that around 500 million years ago these long extinct arthropods were possibly the first animals to colonize the terrestrial environment. Certainly, scientists believe that arthropods, first as detritivores (feeding on dead plants and micro-organisms) and then as carnivores were the first animals to live on land.

Metameric segmentation is where the body is divided into a series of repeated segments, as in a millipede for instance. Each segment then performs all the functions of the body trunk sections, having legs, nerves, breathing apparatus, a unit of digestive tract and all the same organs and tissues. Each segment is in fact a copy of the one before it and the one behind. This is obvious in some arthropods like millipedes, but not so obvious in others. What has happened in the others is called tagmatization. This is where groups of segments become specialized to perform specific functions for the whole body. These groups of segments are called tagmata. Careful dissection and analysis can reveal the underlying form of the original metameric segment in most cases. Some tagmatization occurs in the annelids, i.e. the head with its various appendages. However in the arthropods, it has become far more advanced, reaching its ultimate expression in animals like spiders and barnacles which do not appear to have any segmentation at all to the casual observer.

9–2 Subphylum Chelicerata

● Class Merostomata (Horseshoe Crabs)

The class Merostomata is an ancient class of animals who evolved in the early Cambrian (570 million years ago) and was a dominant life-form in the seas and oceans of the world until the Permian (310 million years ago). Now they are a very small group consisting of only four species in three genera and one order (or subclass) Xiphosura. In the past, the class contained a second order (or subclass) known as Eurypterida, large aquatic animals that looked more like scorpions than their immediate relatives the horseshoe crabs. They have been called sea scorpions though they often lived in freshwater or brackish habitats. Modern horseshoe crabs can occur in large numbers where they still exist. However, all species are considered vulnerable by the IUCN because their populations have often been exploited for many years as animal feed and fertiliser. One species, *Limulus polyphemus* lives off the coast of eastern America, the other three species live in Southeast Asia.

In the breeding season (spring for *L. polyphemus*), horseshoe crabs migrate to certain shallow coastal waters where they collect in large numbers. The male selects a female and cling onto her back. The female then digs a hole about 15 cm deep in the sand in the tidal zone. She lays up 1,000 eggs in this hole and the male releases his sperm onto them so that they can be fertilized. The female then covers the eggs with sand which protects them from the action of the waves while allowing them to be warmed by the sun. The larvae take about 6 weeks to hatch and pass through as many as 16 moults before they reach the adult form. They then have some further growing to do before they reach sexual maturity at three or more years of age. In their early stages, the larvae look a bit like trilobites and are sometimes referred to as "trilobite larvae". This long juvenile period suggests that they are long lived animals but there is little information on just how long they live as adults.

The whole body of the animal is covered, and thus protected by two dorsal plates called a carapace. The carapace is said to bear two compound and two simple eyes. However, the compound eyes are really just a group of simple eyes compressed together (Fig. 9-1). Beneath the horseshoe shaped part of the carapace is the cephalothorax which bears the animals walking and feeding legs. There are five pairs of walking legs, the first of which are sometimes called the pedipalps (feeding legs). The first four pairs of walking legs have pincers and also have their coxae modified to form the gnathobases. The abdomen (or opisthosoma) consists of nine fused segments and is protected by the abdominal dorsal plate which hinges with the cephalothoracic plate. The legs consist of coxa, trochanter, femur, tibia and two tarsal segments. The first pair (of 6) of abdominal legs have the openings of the reproductive organ in them.

The last five of the abdominal legs have gills on them. These gills are called book gills because

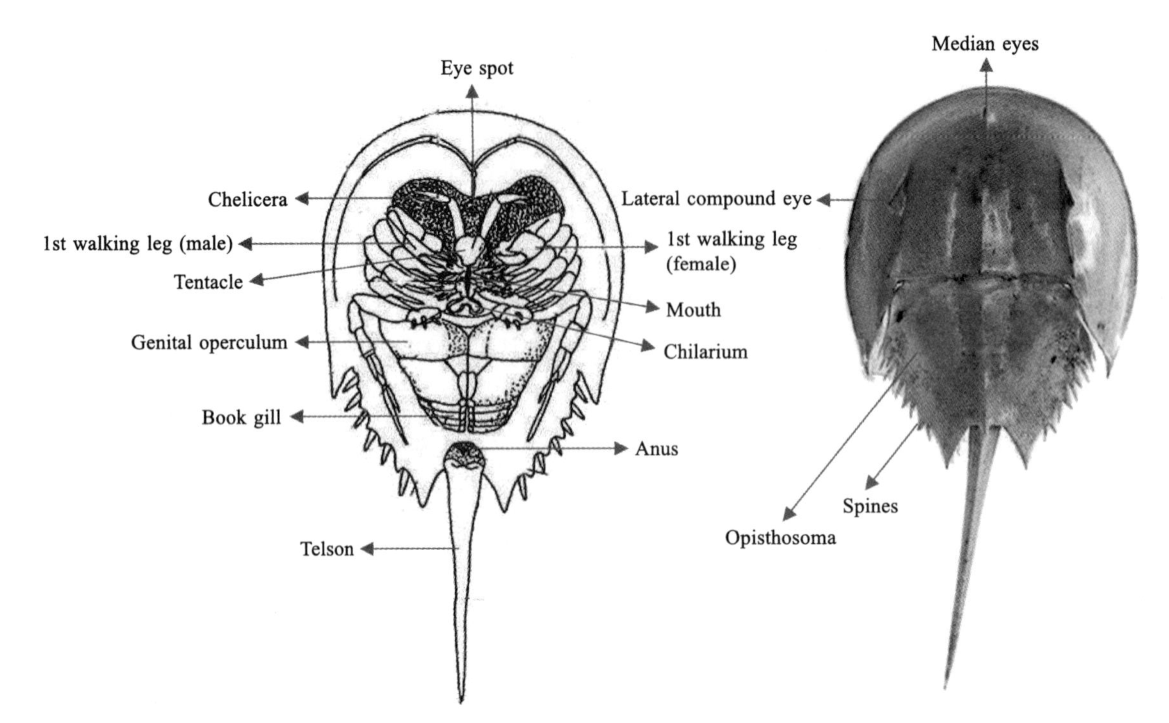

Fig. 9-1 Schematic of horseshoe crabs

each one consists of a series of 150 to 200 flat plates laid one over the other like the pages of a book. These are the organs of gaseous exchange. Horseshoe crabs also have a well-developed circulatory system comprised of a heart and arteries that carry the blood around the body and then into two veins which pass through the gills where carbon dioxide and oxygen are exchanged before the reoxygenated blood returns to the heart. Metabolic wastes are extracted from the blood by two pairs of coxal glands. The glands create urine which passes to a bladder and is excreted through a special pore at the base of the 4th pair of walking legs.

The nervous system consists of a circumoesophageal brain with a vetral nerve cord which gives rise to five sets of ganglia from which nerves spread out to the body. There is also a pair of longitudinal nerve cords.

● **Class Arachnida (Spiders)**

Animals of the class Arachnida are the most familiar of the arthropods outside of the ever-present insects. There are over 80, 000 named species and they are united by the possession of eight legs, chelicerae and pedipalps, and the lack of biting and chewing mouthparts (food is ingested in most cases as a liquid, though some help is often given the food in becoming liquid by the secretion of, or regurgitating onto or into the food of, digestive enzymes from the stomach).

The arachnids are evolved in the sea, but now they are almost entirely terrestrial and have developed several important features to help them survive on the land. These include a waterproof (waxy) exocuticle, internal fertilization, Malpighian tubules as a metabolic excretory system and internal organs for breathing and gaseous exchange (Fig. 9-2).

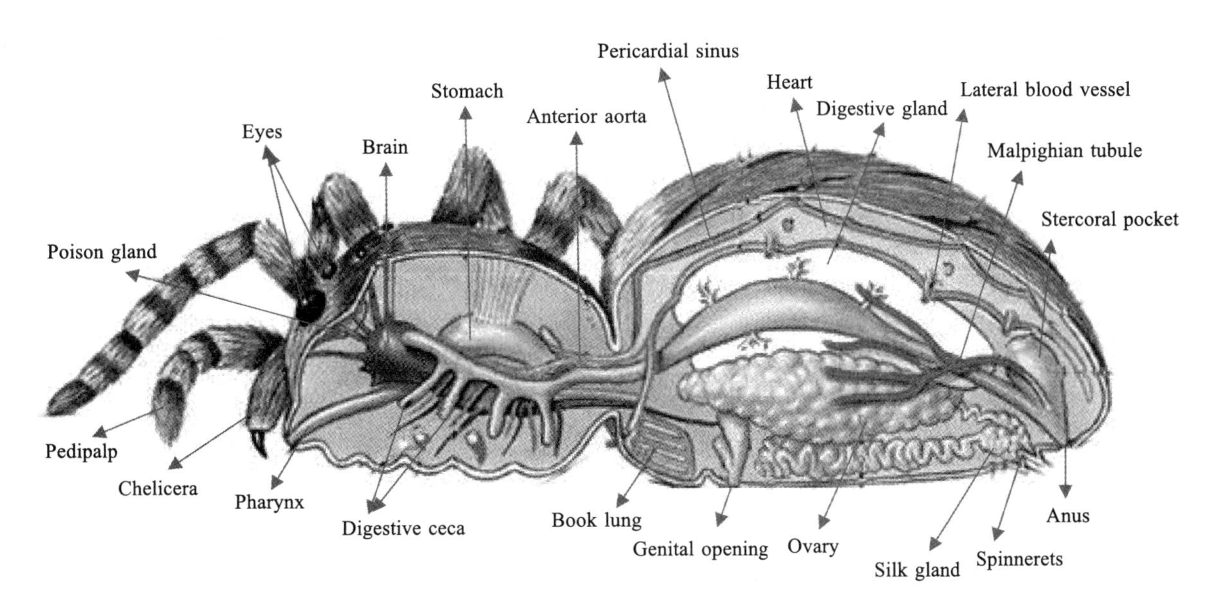

Fig. 9-2 Schematic of arachnids

The class Arachnida is extremely diverse in form and in lifestyles and little more can be said that includes them all. The class Arachnida is divided into 13 subclasses (or orders depending on which classification scheme you are following) of which the Araneae (spiders) and the Scorpiones (scorpions) are the best known.

● **Class Pycnogonida (Sea Spiders)**

Pycnogonids (Fig. 9-3) are odd looking creatures which live in the seas and oceans of the world and normally have four pairs of walking legs, but they may have five or even six pairs in some cases. They have practically no body and a proboscis. They have been relatively little studied and there is a great deal we still do not know about them.

Fig. 9-3 Sea spider (*Sericosura verenae*)

The fossil record of pycnogonids is very meagre. *Paleopantopus maucheri*, with an early Devonian fossil discovered, is one of the few well documented species. Despite this lack of hard evidence, scientists deduce from morphological and embryonic studies that the pycnogonids are an old lineage of animals, though nobody can really put a date on their first appearance.

Pycnogonids are found all over the world, from coastal tropical waters to the poles. They are also found at depths as great as 7,000 m, though they are far more common in shallower waters. They range in size from a few millimetres of legspan to giants with a legspan of 75 cm. As of the late 1990's, there were about 1,000 species known to science divided into 8 families and 86 genera. They are common in the Mediterranean, the Carribean and around the poles, and not that difficult to find in rock pools once you get your eye in.

9–3 Subphylum Crustacea

● **Class Cephalocarida (Small Primitive Shrimps Discovered in 1955)**

The body is elongate, measuring as much as 3 mm in length (Fig. 9-4). It consists of a head, a ten-segmented thorax with legs, and a nine-segmented legless abdomen. At the end of the body is a ramus with two long bristles. The legs are used for locomotion and respiration, as well as for directing food toward the mouth opening. There are two pairs of antennae on the head, small upper jaws, and two pairs of lower jaws, which are practically indistinguishable from the thoracic limbs. The animals lack eyes, a condition related to their burrowing way of life. The female lays eggs into an egg sac located on the last segment of the thorax. Nauplii are hatched from the eggs and become adults only after 18 molts.

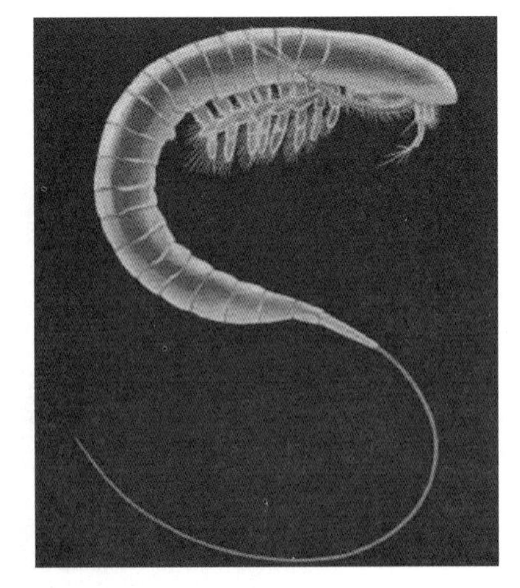

Fig. 9-4 Cephalocaridans

The subclass Cephalocarida was discovered in 1955. Its first identified representative, *Hutchinsoniella macracantha*, was found on the Atlantic coast of the USA. Other species have been found on the eastern and western coasts of North America and near Japan. Three genera, embracing four silt-dwelling species, are known.

● **Class Branchiopoda**

Class Branchiopoda includes small primitive animals with gills on their feet. There are more than 2, 000 described species, and new taxa are still being discovered. Included in this group are fairy shrimps (Anostraca), tadpole shrimps (Notostraca), smooth clam shrimps (Laevicaudata), spiny clam shrimps (Spinicaudata), tropical clam shrimps (Cyclestherida, the so called "large branchiopods"), and the various species of "water-fleas" (Cladocera).

● **Class Ostracoda (Small Animals Which Look Like Miniature Bivalves)**

Ostracods derive their common names from their bivalved shells and small size (and the fact that the name "clam shrimp" is already taken) (Fig. 9-5). The two lateral valves (of calcium carbonate and chitin) are articulated by a dorsal ligament and closed by a single adductor muscle. Segmentation is not clearly evident in the body plan, and the typical crustacean appendages and tagmata are highly modified. There are only three pairs of thoracic segments, and the abdomen is reduced to a pair of caudal furcae. In general, the 1st and 2nd antennae are used for swimming, although ostracods generally remain close to the substrate. The other head appendages bear branchial plates that beat to generate a water current for gas exchange and selective filter-feeding. Ostracods lack compound eyes. Among freshwater species, asexual reproduction is common, and zygotes/eggs may serve as resting stages during harsh environmental conditions. Development is indirect, with a nauplius larval stage.

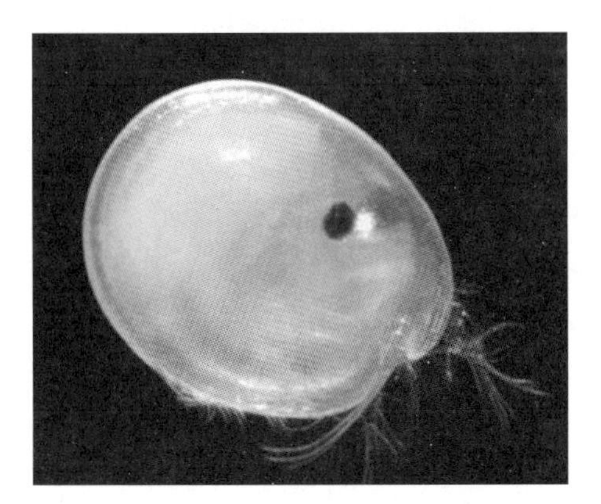

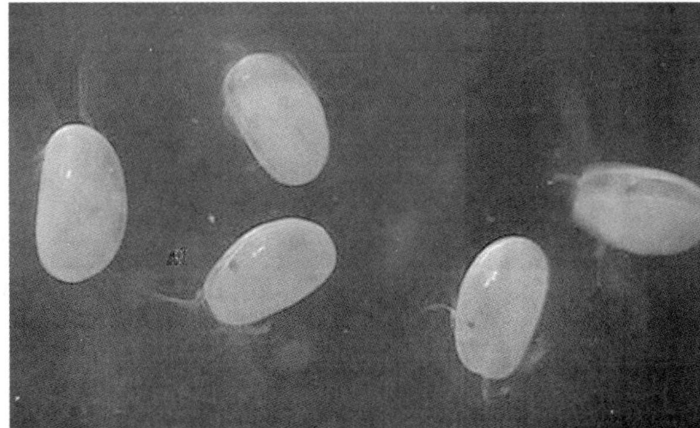

Fig. 9-5 Ostracods

● **Subclass Mystacocarida (Minute Thin Shrimp-like Creatures)**

Mystacocarida is a subclass of crustaceans, which form part of the meiobenthos. They are less than 1 mm long, and live interstitially in the intertidal zones of sandy beaches (Fig. 9-6).

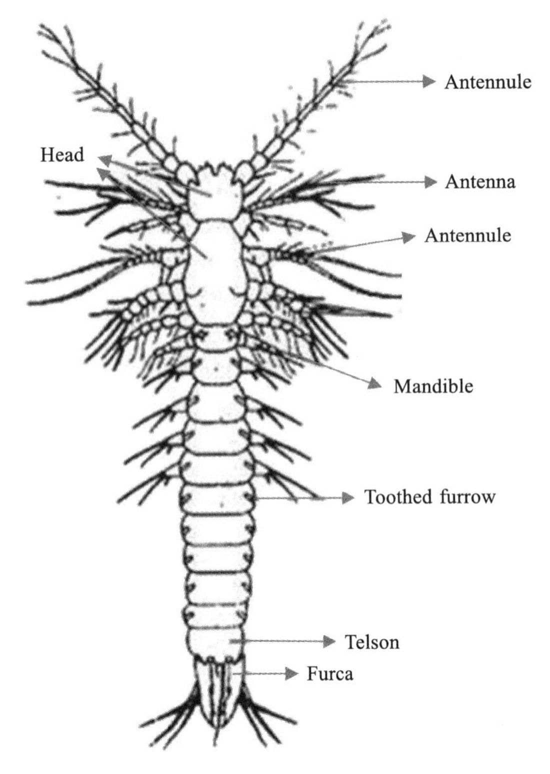

Fig. 9-6 ***Derocheilocaris***

● Class Copepoda (Important Small Crustaceans Such as *Cyclops*)

Copepods (Fig. 9-7) are found in great abundance in waters all over the world, particularly in the oceans. They have a few common names such as sea insects, or fish lice, the latter term being a reference to the parasitic lifestyle of some species. Over 30 species inhabit arctic lakes and ponds, including species belonging to the genera *Limnocalanus*, *Diaptomus*, and *Cyclops*. Although there is tremendous variation in body form, most copepods have elongated, segmented bodies, a single eye and two long antennae. Each segment of the thorax bears a pair of swimming legs and there are two appendages that project from the final body segment. The largest copepods attain lengths greater than 3 mm, although most arctic freshwater species are less than 1 mm. *Lernaeopoda salmonea* is a parasite that occurs commonly on the gills and mouth parts of Arctic charr. It looks like a white, formless blob, less than 1 cm in length.

Fig. 9-7 Copepods

Because freshwater lakes in the Arctic contain very little plant matter, compared to other regions, their water is extremely clear and easily penetrated by ultraviolet (UV) light. This presents a problem for many freshwater organisms, including copepods, in that they are at risk of sunburn. Many arctic planktonic animals are pigmented, an adaptation that reduces UV damage. Arctic copepods, such as *Diaptomus minutus*, which live throughout the water column are often red or orange in color.

Copepods, because of their abundance, are often an important food source for other invertebrates and juvenile fish. Most arctic freshwater copepods filter algae and bacteria from the water or prey upon small invertebrates. As a result, they are a vital link in the food web; they typically consume plants and small animals, and provide an important food supply for larger animals. All species of freshwater copepods reproduce sexually, with males and females coming together to combine eggs and sperms. Typically, it is the eggs that survive in a dormant state through the winter, developing into adults the following spring. Arctic populations of *Limnocalanus macrurus* develop into adults from July to August, and mate between October and November. Their eggs are released into the water, sink to the bottom, and hatch in about one month. The young develop throughout the winter months.

● **Subclass Branchiura (Small Blood Sucking Ectoparasites)**

Crustaceans of the subclass Branchiura, or fishlice, are distinctive ectoparasites of marine and freshwater fishes. There are more than 120 freshwater species (one family), only 18 species (one genus) occur in in-land habitats in North America.

Branchiurans are generally in the millimeter to centimeter size range. A carapace covers the entire cephalothorax, bearing appendages adapted for a parasitic lifestyle. The 1st and 2nd antennae as well as the maxillules are used for attachment to their host, and the mandibles are piercing organs. There are four pairs of biramous swimming legs, and the abdomen is reduced. Branchiurans have compound eyes (Fig. 9-8).

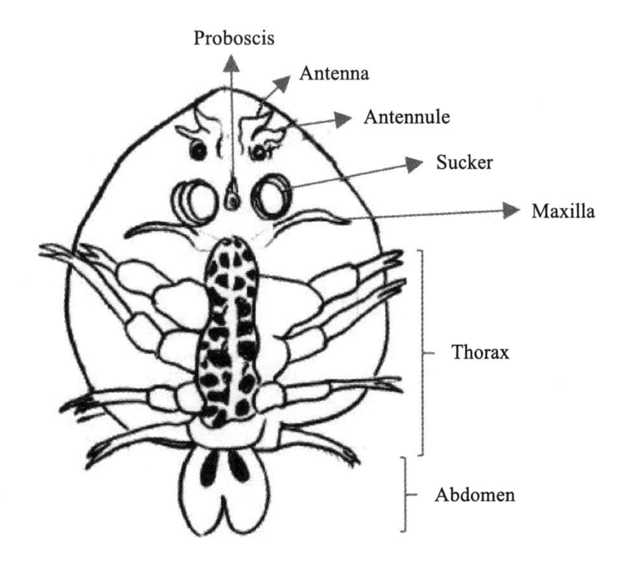

Fig. 9-8 Schematic of *Argulus foliaceus*

● **Subclass Cirripeda (Barnacles)**

Barnacle, also called cirripede, is any of more than 1,000 predominantly marine crustaceans of the subclass Cirripedia highly modified for sedentary life. There are about 850 free-living species (all marine) and about 260 species that are internal parasites of crabs and other crustaceans.

As adults, typical barnacles are covered with calcareous plates and are cemented, head down, to rocks, pilings, ships' hulls, driftwood, or seaweed, or to the bodies of larger sea creatures, from clams to whales. They trap tiny particles of food by means of cirri—feathery retractile organs formed by metamorphosis of certain of their swimming legs (Fig. 9-9).

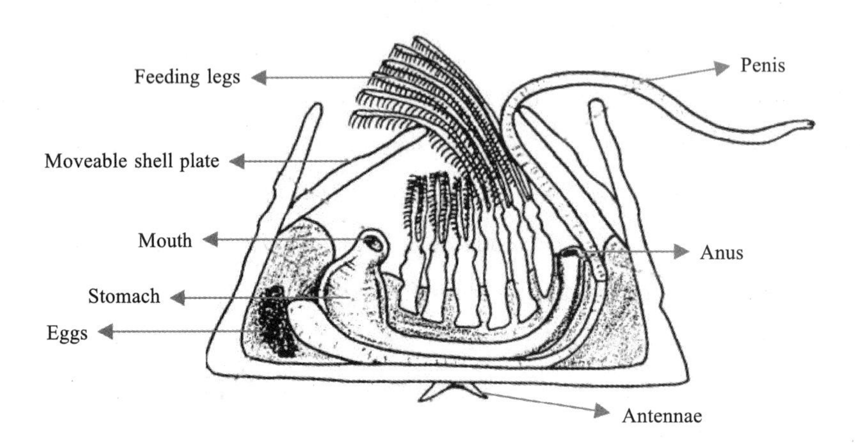

Fig. 9-9 Barnacle

● **Class Malacostraca (75% of all Crustaceans, Crabs, Lobsters, Shrimps and Prawns as well as Woodlice)**

The class Malacostraca contains the largest number of crustaceans species (Fig. 9-10). Malacostracans are made up of 16 orders and 6 suborders. Most malacostracans live in the marine environments, but a few are found in the freshwater and terrestrial environments. They occupy all three dimensions in the water column, such as crawling on top of the sediment, burrowing in the substrate, or swimming with the oceans' currents. Most malacostracans are predatory by scavenging, but few are detritus feeders, filter feeders, parasites, and herbivores. In the aquatic members, the general trend is to use the gills for the site of ammonia excretion. Fertilization is internal and sexual. The female carries the eggs under the

Fig. 9-10 Malacostracans

carapace in a brood pouch until they are ready to be released as larvae. The largest order is the Decapoda and contains about one quarter of all the species of crustaceans. These are the familiar lobsters, shrimps, and crabs. Three pairs of the first thoracic appendages are modified as maxillipeds, which are used for feeding. The remaining five pairs of legs are used for locomotion, and the first pair is usually modified as a large claw or cheliped. The second largest group is the infraorder Brachyura, sometimes called "true" crabs. The folding of the abdomen beneath the thorax allows for the center of gravity to shift forward underneath the legs, thus providing better mobility. The remaining orders represent the smaller pelagic and sedimentary species.

The shrimp has 19 separate sections of the body. Two main segments make up the body of a freshwater or marine shrimp. The first part is the upper portion of the shrimp, referred to as the cephalothorax. The cephalothorax includes the head and the thorax or pereon region of the shrimp. It is covered by a protective plating system called the carapace.

(1) The Circulatory System of Dwarf Shrimp

The circulatory system of shrimp is open. It means that the entire abdominal cavity of the animal is filled with blood. The heart contracts to send the oxygenated blood to the different parts of the body through eight arteries (Fig. 9-11).

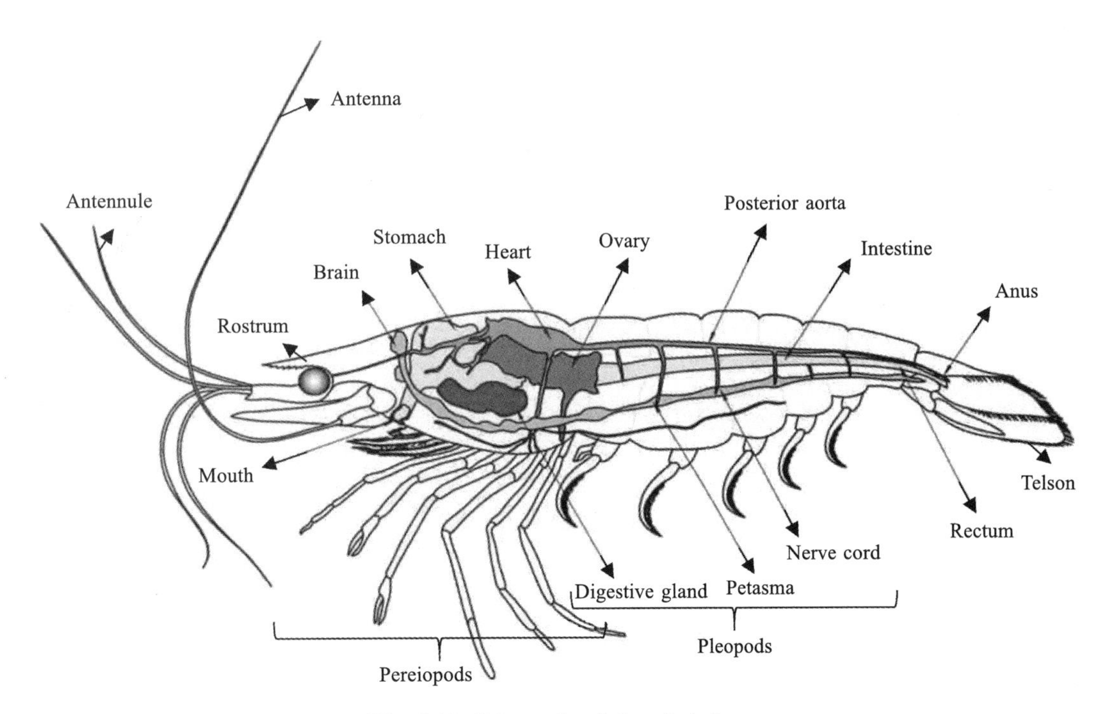

Fig. 9-11 Schematic of dwarf shrimp

(2) The Respiratory System of Dwarf Shrimp

Dwarf shrimp breaths with the help of gills. There are several different morphological types of the gills: dendrobranchie, tricobranchie and fillobranchie. For example, the *Caridina* and *Neocaridina* have fillobranchie.

(3) The Digestive System of Dwarf Shrimp

The digestive system of shrimp can be divided into two main parts:

Alimentary canal. It consists of a mouth, buccal cavity, stomach, rectum, and anus.

Digestive gland (hepatopancreas). The gland functions as a liver, pancreas, and intestine for the shrimp (all in one). Additionally, it absorbs digested nutrients and can store them for some time. Therefore, digestive gland has two functions—digestion and storage.

(4) The Reproductive System of Dwarf Shrimp

In male shrimp, testes (paired and soft organs) produce spermatids that are transported to and accumulate in the vas deferens. The testes are located in the neck area beneath the heart. Each of them includes numerous tiny tubes, called seminiferous tubules. Each tubule has an inner lining of a single layer of epithelial cell which transforms into spermatozoa.

In female sexual mature shrimp, there are paired ovaries, in which the eggs are produced. It is easy to see them beneath the heart if the shrimp is transparent. Aquarists usually call the paired ovaries "saddle". This is the only indicator of shrimp gender, which can guarantee the difference between males and females. The matured eggs remain near the margin and the immature eggs occupy the center.

The reproductive process of dwarf shrimp is shown in Fig. 9-12.

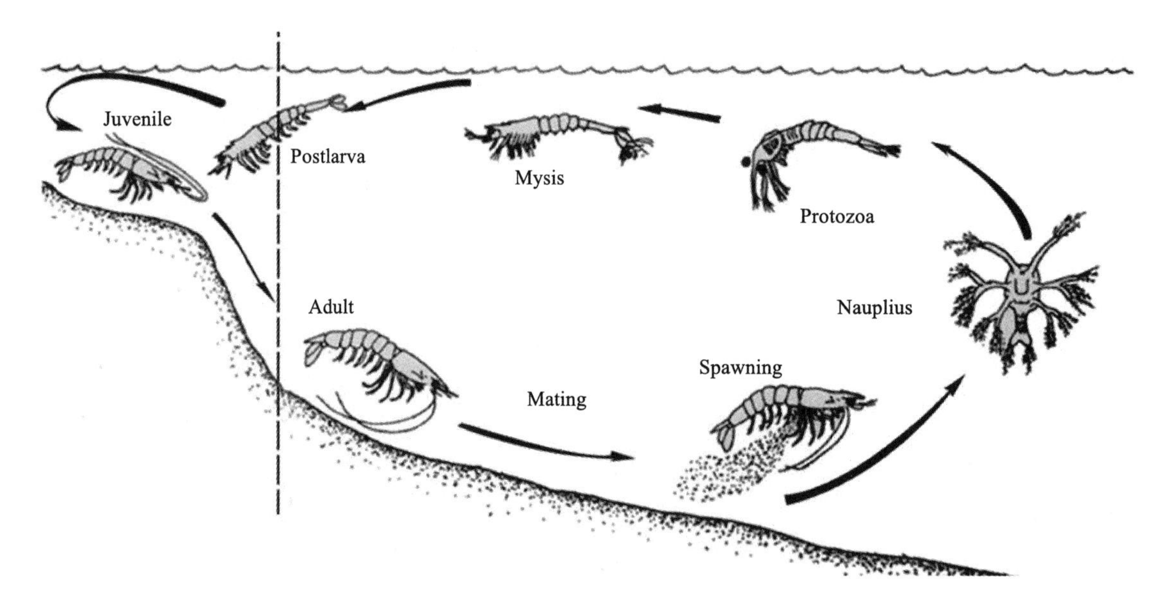

Fig. 9-12 Reproductive process of dwarf shrimp

9-4 Subphylum Uniramia

Uniramians have a body in three parts, and a thorax in three segments.

● **Class Chilopoda (Centipedes)**

Centipede legs extend visibly from the body, with the final pairs of legs trailing behind it. This allows them to run quite fast, either in pursuit of prey or in flight from predators. Centipedes have just one pair of legs per body segment, a key distinction from millipedes (Fig. 9-13).

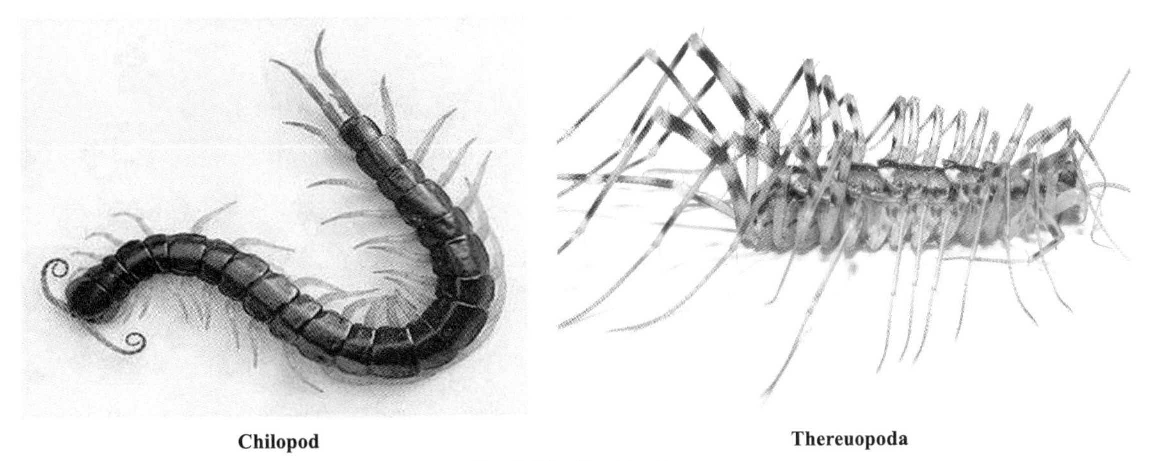

Chilopod Thereuopoda

Fig. 9-13 Centipedes

The centipede body is long and flattened, with a long pair of antennae protruding from the head. A modified pair of front legs function as fangs used to inject venom and immobilize prey.

Centipedes prey on insects and other small animals. Some species also scavenge on dead or decaying plants or animals. Giant centipedes, which inhabit South America, feed on much larger animals, including mice, frogs, and snakes.

While house centipedes may be creepy to find in the home, you might want to think twice about harming them. House centipedes feed on insects, including the egg cases of cockroaches.

● **Class Diplopoda (Millipedes)**

Members of class Diplopoda are more commonly known as millipedes (Fig. 9-14), and they are found throughout the world, especially in damp, wooded areas, but also sometimes in grasslands and other environments. The word "millipede" comes from combining "milli" which means "one thousand", and "pede", which means "feet", so millipedes are commonly believed to all have a thousand feet. However, although they do have lots of legs, no millipede has quite that many. Most usually have only

Fig. 9-14 Millipedes

100–300 legs, although some species can have up to 750 legs.

Millipedes play an important role in the ecosystem, and they have been incredibly successfully adapting to life throughout the world. In fact, they are believed to be one of the very first land animals that evolved, first appearing on earth over 400 million years ago.

● **Class Symphyla (Small Centipedes-Like Creatures Which Live in Leaf Litter)**

Members of class Symphyla are rapid runners that range in length from 2.5 to 10 cm. The class includes some 160 species. They are mainly scavengers on decayed vegetation, but one species, *Scutigerella immaculata*, is a serious pest of certain crops. Symphylans have 12 pairs of legs and resemble the centipedes (Fig. 9-15).

Fig. 9-15 Symphylans

● **Class Pauropoda (Small Soft-Bodied Animals of the Forest Floor)**

There are about 500 known species belonging to class Pauropoda. Pauropods are soft-bodied, small (0.5–2.0 mm long), soil-inhabiting arthropods that are distributed worldwide. They are elongated and have 9–11 pairs of legs, but they have no trachea and no heart (Fig. 9-16).

Fig. 9-16 Pauropods

● **Class Insecta (Insects)**

The earth is literally crawling and buzzing with insects, otherwise known as members of the class Insecta. It would be hard to get through the day without a close encounter with a six-legged species.

Insects are so prolific that added together they outnumber all other life forms combined. They are found on land, in the air, and in the sea. Their diversity is astounding. From the minuscule pesky flea to the majestic monarch butterfly, there are around a million different species. In spite of their vast differences, all insects have certain traits in common (Fig. 9-17).

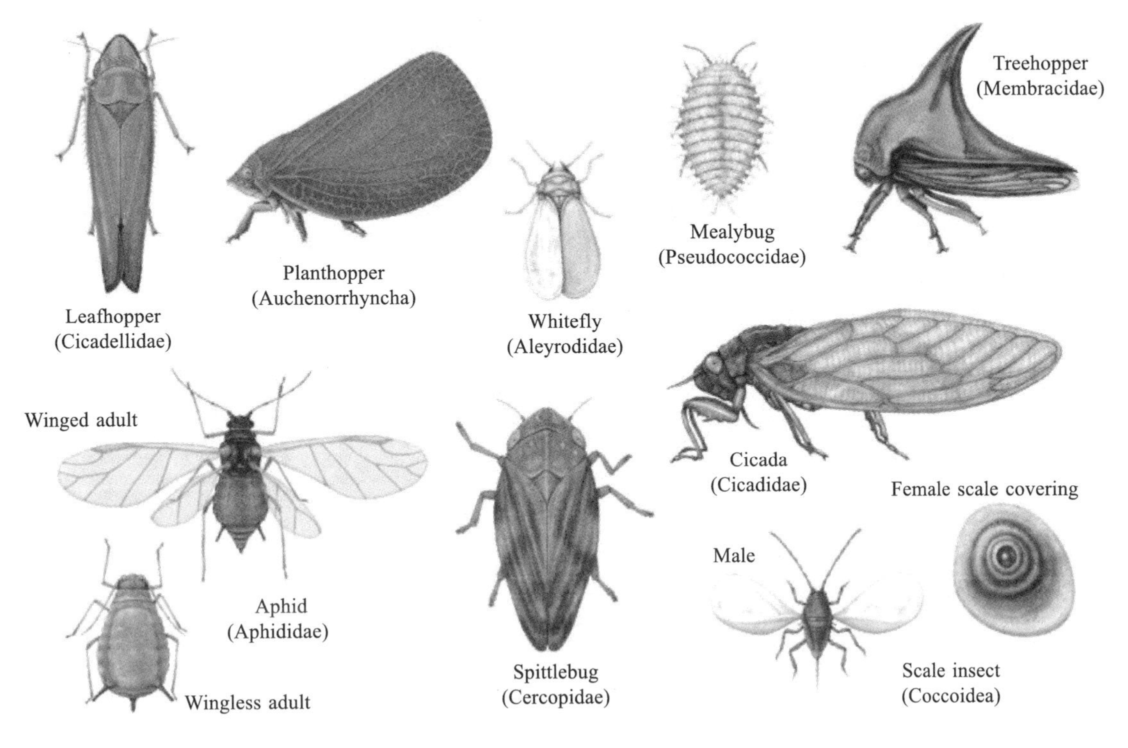

Fig. 9-17 Insects

Insects also show huge variety in shape and form. Almost the only condition their group does not attain is very large body size. A number of features, however, are shared by most kinds of living insects. In addition to the general characteristics of uniramians, these include a body composed of three tagmata, a head, thorax, and abodmen; a pair of relatively large compound eyes and usually three ocelli located on the head; a pair of antennae, also on the head; mouthparts consisting of a labrum, a pair of mandibles, a pair of maxillae, a labium, and a tonguelike hypopharynx; two pairs of wings, derived from outgrowths of the body wall (unlike any vertebrate wings); and three pairs of walking legs (Fig. 9-18).

Insects have a complete and complex digestive tract. Their mouthparts are especially variable, often complexly related to their feeding habits (Fig. 9-19). Insects "breathe" through a tracheal system, with external openings called spiracles and increasingly finely branched tubules that carry gases right to the metabolizing tissues. Aquatic forms may exchange gases through the body wall or they may have various kinds of gills. Excretion of nitrogenous waste takes place through Malpighian tubules (Fig. 9-20). The nervous system of insects is complex, including a number of ganglia and a ventral, double nerve cord. The ganglia are largely independent in their functioning; for example, an isolated thorax is

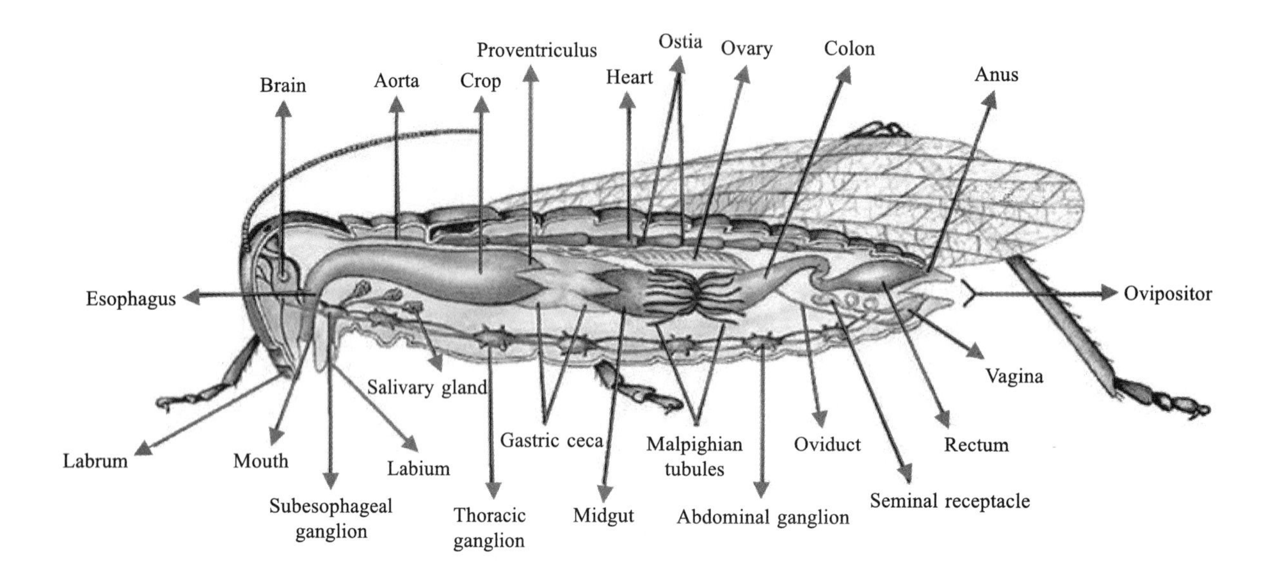

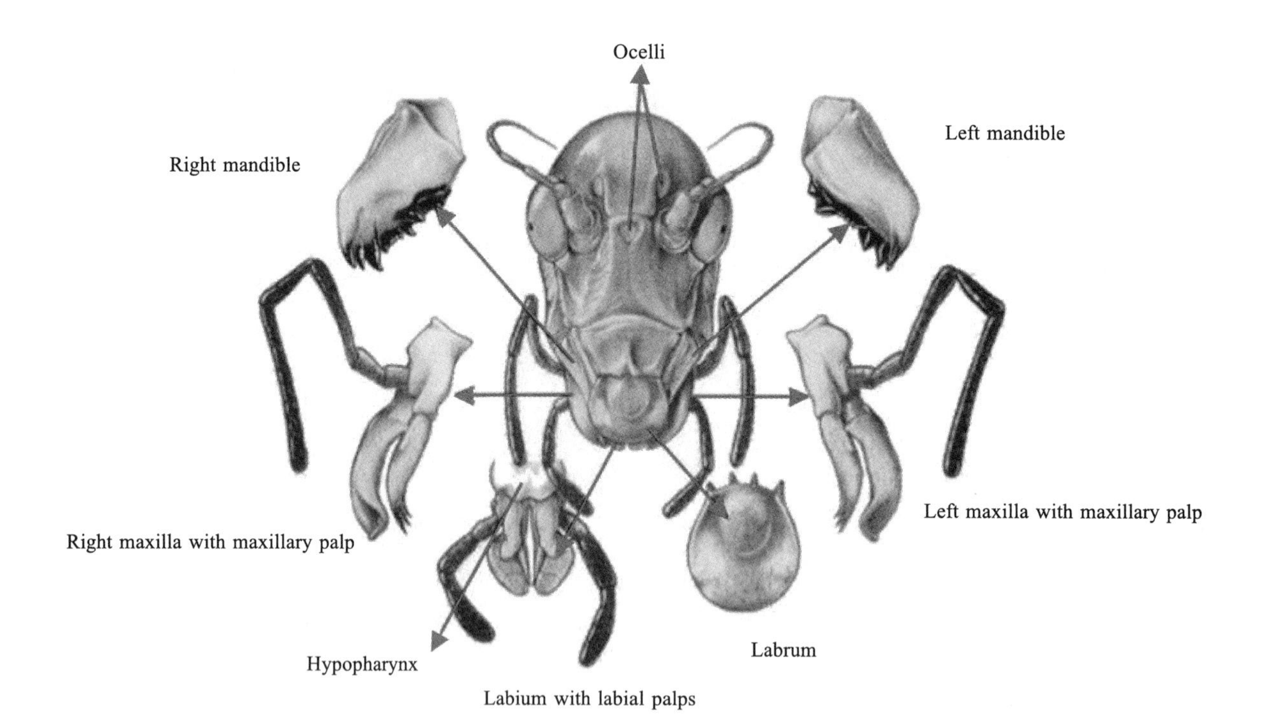

Fig. 9-18 Schematic of grasshopper

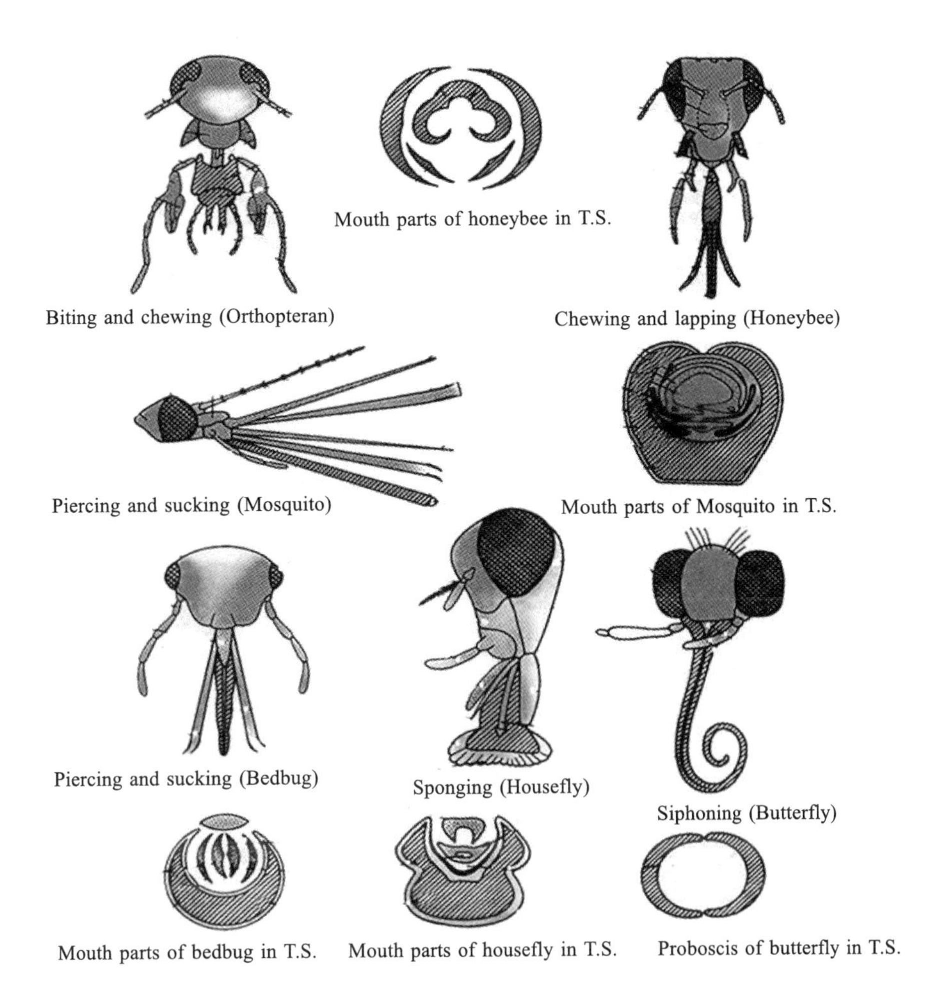

T.S.: transverse section

Fig. 9-19 Mouthparts of insects

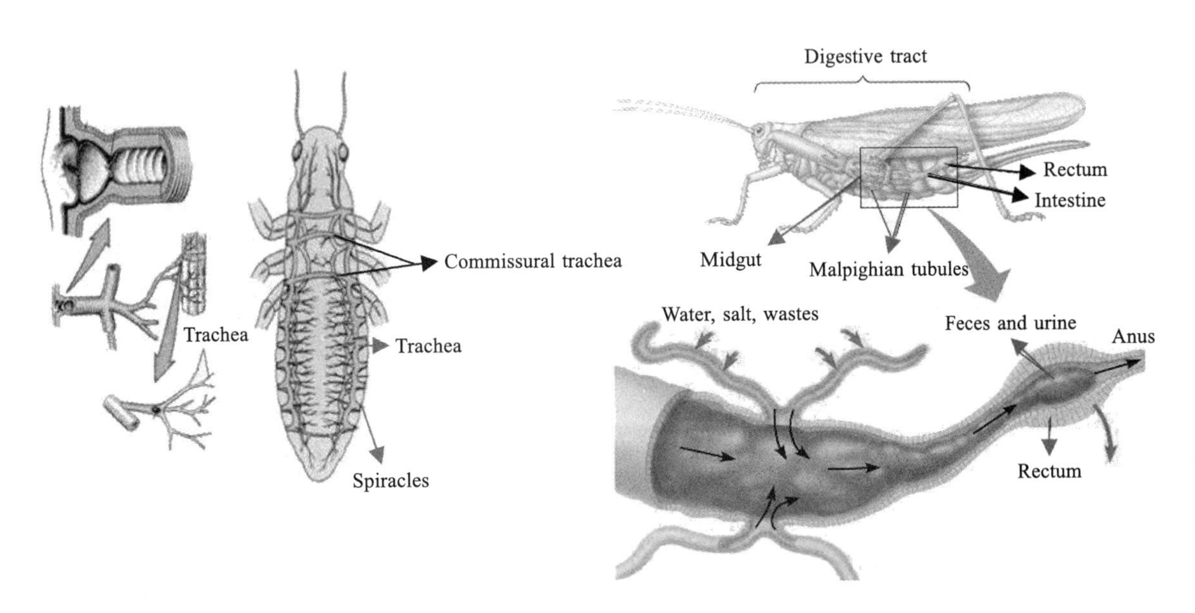

Fig. 9-20 Tracheal system and Malpighian tubules

capable of walking. Yet ganglia also use sensory output. A grasshopper with one wing removed can correct for this loss and maintain flight, using sensory input from its brain. Sense organs are complex and acute. In addition to ocelli and compound eyes, some insects are quite sensitive to sounds, and their chemoreceptive abilities are astounding.

Insects are dioecious and fertilization is internal in most. The ways in which mating is accomplished, however, are incredibly variable; study of this variability by evolutionary biologists has greatly advanced our understanding of the evolution of behavior, social evolution, and traits such as number, size of young and patterns of investment in them. Reproduction by insects often involves a male locating a receptive female through chemicals (pheromones) released by the female. In most species, females store the sperm in a special receptacle in their abdomens; even species that lay huge numbers of eggs (in honeybees, for example, the number may be over one million), females mate only once and rely on sperm stored during that mating for the rest of their lives.

The manner in which growth is accomplished is an especially important characteristic of insects. In some, hatching eggs produce miniature adults, which to grow must shed their exoskeleton in a process called ecdysis. In almost 90% of insect species, however, newly hatched young are completely different in appearance from adults. These larval forms usually live in different habitats, eat different foods, and assume a body form completely different from that of their parents. The larva feeds and grows, molting its skin periodically. At some point larval growth is completed, the larva stops feeding and builds a case or cocoon around itself. In this nonfeeding condition it is called a pupa or chrysalis. While so encased, the larva undergoes a complete transformation or "metamorphosis" of its body form, and a fully-formed adult emerges. Insects that experience this sort of complete change are called "holometabolous". Other species undergo a more gradual process, in which the newly hatched young are more similar to the adult but are small in size, lack wings, are sexually immature, and may differ in other relatively minor ways as well. The young in these insects are called nymphs, and the lifestyle is referred to as "hemimetabolous" (Fig. 9-21).

● **Insect Classification**

The class Insecta can be further divided into 29–32 orders depending upon the classification system used. Species in each order share unique characteristics that set them apart from other insects, yet have the same traits that all insects share. Let's outline a few of the more common orders containing familiar insects.

(1) Coleoptera (Beetles)

Coleoptera is the largest order of any animal with approximately 370,000 known species. Beetles make up 25% of all known plants and animals, and are found in every ecosystem except salt water and polar ice caps. Their exoskeleton is very hard. They have biting mouthparts and two pairs of wings, one for flying and one for a protective covering. Larvae are considered "grubs". Life cycle is complete metamorphosis (egg, larva, pupa, adult) (Fig. 9-22).

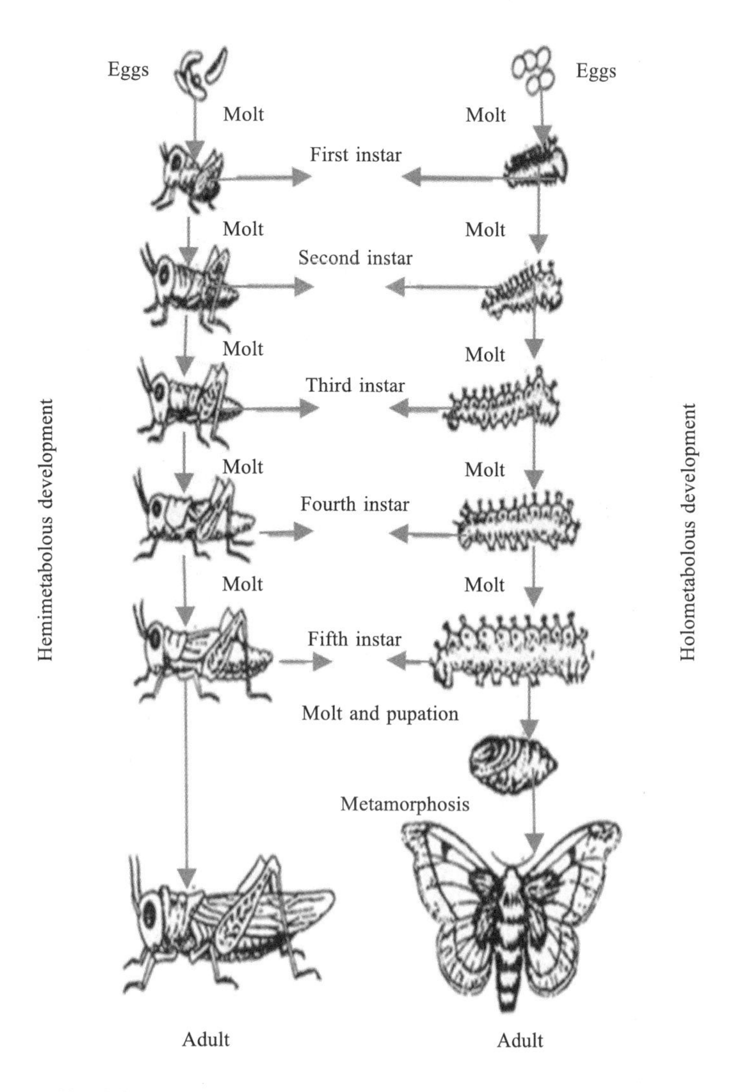

Fig. 9-21 Holometabolous and hemimetabolous development

Fig. 9-22 Beetles

(2) Lepidoptera (Butterflies and Moths)

Lepidoptera is the second largest animal order with over 160,000 known species. They have large wings covered in tiny scales. Their wings are often brilliantly colored or patterned. Adults feed on nectar gathered by a long tube that extends from their mouths called a proboscis. They "taste" with their feet. Life cycle is complete metamorphosis (egg, larva, pupa, adult) (Fig. 9-23).

Fig. 9-23 Butterflies and moths

For the most part, moths are nocturnal. They fly and feed at night. Butterflies, on the other hand, are diurnal, or active during the day. Butterflies and moths also tend to hold their wings differently when they rest. You'll often see moths with their wings draped down their backs or spread out to the side. Butterflies and moths also pupate, or become adults, differently. Both go through their metamorphosis in a chrysalis, or protective shell. Moths, however, often spin a silk cocoon around their chrysalis, sometimes camouflaging it with leaves or debris. Butterflies' antennae are wider at the tips—their ends look like little clubs. Moths' antennae are often feathery. Sometimes, they're thin like butterflies' antennae, but without the clubs. Most of the time, butterflies' wings display more vivid colors than moths' wings do. Often, moths' bodies are plumper and fuzzier than butterflies' bodies.

(3) Hymenoptera (Ants, Bees, and Wasps)

Their characteristics include: two pairs of wings; bees visit flowers to collect pollen and nectar; may have chewing mouthparts or tubes to collect nectar from flowers; can smell with their antennae; wasps visit flowers to prey on smaller insects (often pests); well-developed compound eyes; can be social or solitary; social species have a highly developed social structure; life cycle is complete metamorphosis (egg, larva, pupa, adult) (Fig. 9-24).

Fig. 9-24 Ants (left), bees (middle), and wasps

(4) Diptera (Flies, Gnats, and Mosquitoes)

Their characteristics include: two pairs of wings; forewings are used for flight and hindwings are used for balance; well-developed compound eyes; mouthparts designed to suck or pierce, not bite; feed on liquids; life cycle is complete metamorphosis (egg, larva, pupa, adult); considered pests and some are disease carriers (Fig. 9-25).

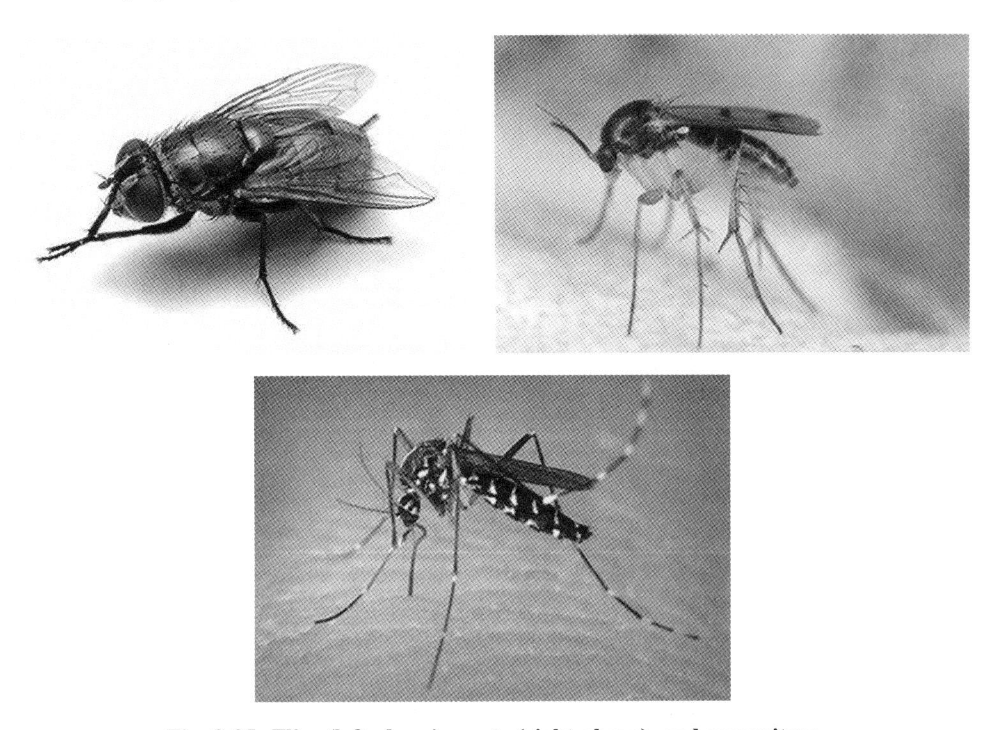

Fig. 9-25 Flies (left above), gnats (right above), and mosquitoes

(5) Orthoptera (Grasshoppers and Crickets)

Their characteristics include: well-developed mandibles and can bite; two pairs of wings, hindwings for flight and forewings are leathery wing protectors; large hind legs allow them to jump; can "sing" or "chirp" by rubbing parts of their bodies together; life cycle is incomplete metamorphosis (egg, nymph, adult); grasshoppers eat plants, and crickets hunt smaller insects (Fig. 9-26).

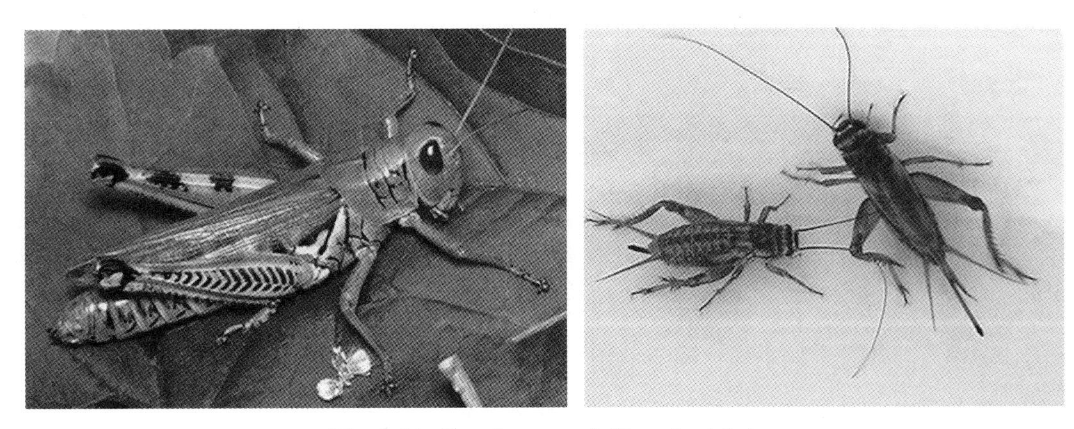

Fig. 9-26 Grasshoppers (left) and crickets

(6) Hemiptera ("True Bugs", Cicadas, Aphids)

Their characteristics include: piercing mouthparts to suck juice from plants or animals; some species have two pairs of wings, some have one pair, and others have no wings; life cycle is incomplete metamorphosis (egg, nymph, adult); considered agricultural pests (Fig. 9-27).

Fig. 9-27 Cicadas (left) and aphids

(7) Odonata (Dragonflies and Damselflies)

Their characteristics include: relatively large insects with long bodies; two pairs of similarly sized wings that are transparent and veined; large heads and large compound eyes; very small antennae; life cycle is incomplete metamorphosis (egg, nymph, adult); adults eat smaller insects (Fig. 9-28).

Fig. 9-28 Damselflies (left) and dragonflies

9-5 学习要点

● 本门动物身体异律分节，具成对而分节的附肢、几丁质的外骨骼和混合体腔hemocoele的生物学意义；肌肉、循环、呼吸诸器官系统结构与功能之间的高度统一关系。

● 同律分节与异律分节的不同；环节动物、软体动物和节肢动物血窦来源上的不同。

● 节肢动物门分为有鳃亚门、有螯肢亚门和有气管亚门。

● 甲壳纲代表动物——中国对虾*Penaeus chinensis*的形态构造，头胸甲和附肢（单肢型和双肢型，螯肢）的识别；对虾的4种鳃的着生部位及形状的不同；对虾的生活习性、个体发育、洄游现象、经济价值等；甲壳纲常见的亚纲以及软甲亚纲重要的类群；蚤*Daphnia*、剑水蚤*Cyclops*、藤壶*Balanus*、沼虾*Macrobrachium*、绒螯蟹*Eriocheir*、梭子蟹*Portunus*的分类地位、经济意义和分布。

● 肢口纲的基本特征、书鳃、生活习性、分布和经济意义。

● 蜘蛛的基本形态和结构、纺织腺；蛛形纲重要的类群（蝎目、蜘蛛目和蜱螨目）。

● 昆虫纲代表动物——东亚飞蝗*Locusta migratoria manilensis*的主要形态结构特征；昆虫的触角、口器、足和翅的结构特征；昆虫的不同种类由于适应不同生活环境而产生的各种形态结构变化类型；蝗虫的生殖和发育；昆虫的滞育、拟态和多态等概念；昆虫的主要常见种类和水生种类的分类地位及与人类的关系。

● 同律分节与异律分节。

● 节肢动物血窦的来源与结构特征。

● 节肢动物结构与功能的一致关系。

● 对虾19对附肢的构造、功能分化和数量，以及鳃的类型。

● 昆虫的蜕皮和变态，以及变态类型。

9–6 巩固测验

【名词】

异律分节、完全变态、不完全变态、马氏管、绿腺、混合体腔、卵胎生、基节腺、书肺、羽化

【选择】

1. 以下不属于节肢动物繁盛的原因是（　　）。

A. 几丁质外骨骼　　B. 变态现象　　C. 繁殖能力强　　D. 身体同律分节

2. 以下动物属于昆虫纲的是（　　）。

A. 黑寡妇　　B. 蝎子　　C. 人虱　　D. 水蛭

3. “小荷才露尖尖角，早有蜻蜓立上头。”诗中所写动物的触角类型是（　　）。

A. 丝状　　B. 刚毛状　　C. 锯齿状　　D. 棒状

【填空】

1. 蟹、鲎、蜘蛛、蝴蝶分别主要用（　　　　）、（　　　　）、（　　　　）、（　　　　）进行呼吸。

2. 在节肢动物中，（　　　　）、（　　　　）和（　　　　）均为与（　　　　）肾管同源的腺体结构，可排泄代谢废物。

3. 昆虫的变态有（　　　　）、（　　　　）、（　　　　）、不全变态和完全变态五种基本类型。

【简答】

1. 节肢动物的主要特征有哪些?

2. 昆虫的口器由哪几部分组成？举例说明昆虫主要的口器类型。

3. 对虾附肢的名称与功能分别是什么?

4. 昆虫的呼吸系统为何是动物界中高效的呼吸系统?

5. 举例说明昆虫发育的不同类型。

6. 蝶和蛾有何区别?

Chapter 10 Phylum Echinodermata

10-1 Characteristics of Phylum Echinodermata

- Adult echinoderms exhibit pentaradial symmetry and have a calcareous endoskeleton made of ossicles, although the early larval stages of all echinoderms have bilateral symmetry. The endoskeleton is developed by epidermal cells and may possess pigment cells, giving vivid colors to these animals, as well as cells laden with toxins. Gonads are present in each arm. In echinoderms like sea stars, every arm bears two rows of tube feet on the oral side. These tube feet help in attachment to the substratum. These animals possess a true coelom that is modified into a unique circulatory system called water vascular system. An interesting feature of these animals is their power to regenerate, even when over 75 percent of their body mass is lost.
- Echinoderms possess a unique ambulacral or water vascular system, consisting of a central ring canal and radial canals that extend along each arm (Fig. 10-1). Water circulates through these structures and facilitates gaseous exchange as well as nutrition, predation, and locomotion. The water vascular system also projects from holes in the skeleton in the form of tube feet. These tube feet can expand or contract based on the volume of water present in the system of that arm. By using hydrostatic pressure, the animal can either protrude or retract the tube feet. Water enters the madreporite on the aboral side of the echinoderm. From there, it passes into the stone canal, which moves water into the ring canal. The ring canal connects the radial canals (there are five in a pentaradial animal), and the radial canals move water into the ampullae, which have tube feet through which the water moves. By moving water through the unique water vascular system, the echinoderm can move and force open mollusk shells during feeding.
- The nervous system in these animals is a relatively simple structure with a nerve ring at the center and five radial nerves extending outward along the arms. Structures analogous to a brain or derived from fusion of ganglia are not present in these animals.

- Podocytes, cells specialized for ultrafiltration of bodily fluids, are present near the center of echinoderms. These podocytes are connected by an internal system of canals to an opening called the madreporite.
- Echinoderms are sexually dimorphic and release their eggs and sperms into water; fertilization is external. In some species, the larvae divide asexually and multiply before they reach sexual maturity. Echinoderms may also reproduce asexually, as well as regenerate body parts lost in trauma.
- This phylum is divided into five extant classes: Asteroidea (sea stars), Ophiuroidea (brittle stars), Echinoidea (sea urchins and sand dollars), Crinoidea (sea lilies or feather stars), and Holothuroidea (sea cucumbers).

Echinodermata are so named owing to their spiny skin (from the Greek "echinos" meaning "spiny" and "dermos" meaning "skin"), and this phylum is a collection of about 7,000 described living species. Echinodermata are exclusively marine organisms. Sea stars, sea cucumbers, sea urchins, sand dollars, and brittle stars are all examples of echinoderms. To date, no freshwater or terrestrial echinoderms are known.

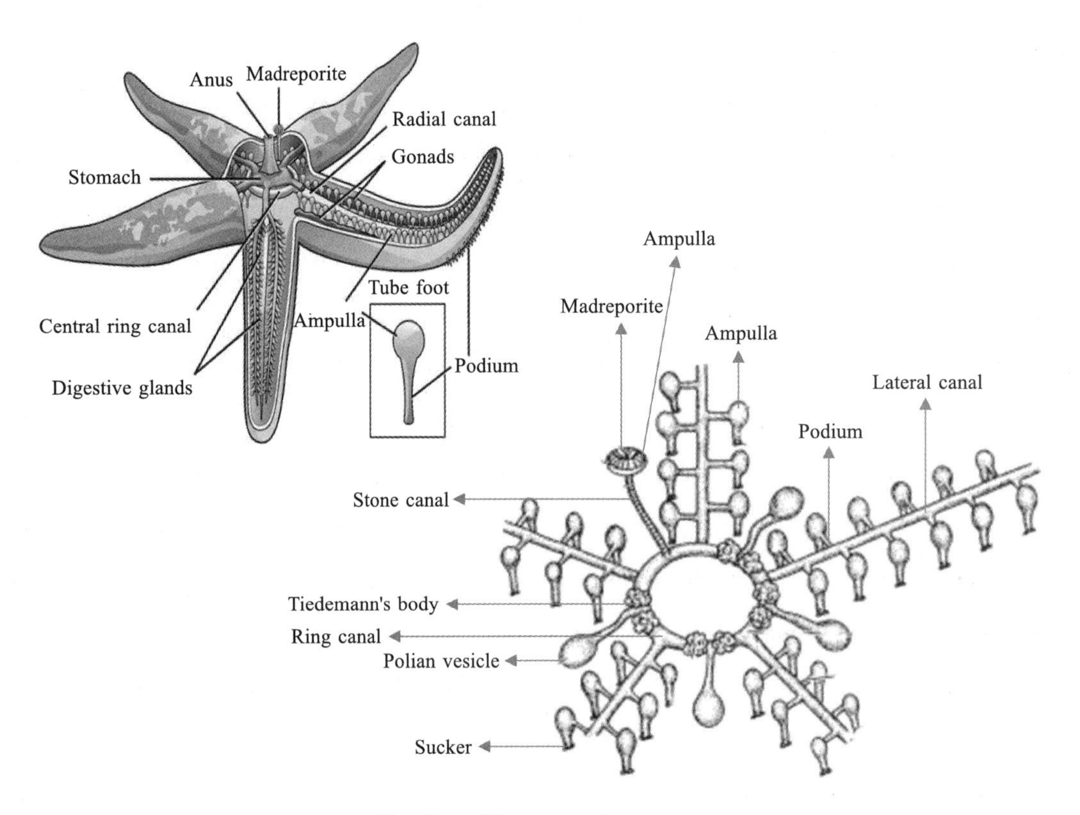

Fig. 10-1 Water vascular system

10–2 Class 1: Asteroidea

Their characteristics include: body is flattened star shaped with five arms; they possess tube feet with a sucker; presence of calcareous plates and movable spines; respiratory organ is papulae.

Examples: *Asterias*, *Astropecten*, *Zoroaster*, *Oreaster* (Fig. 10-2).

Fig. 10-2 Sea stars

10–3 Class 2: Ophiuroidea

Their characteristics include: body is flat with pentamerous disc; they possess a long arm which is sharply demarcated from the central disc; they possess tube feet without sucker; anus and intestine are absent; respiratory organ is bursae.

Examples: *Ophiodera*, *Ophiothrix*, *Astrophyton*, *Amphiura*, etc (Fig. 10-3).

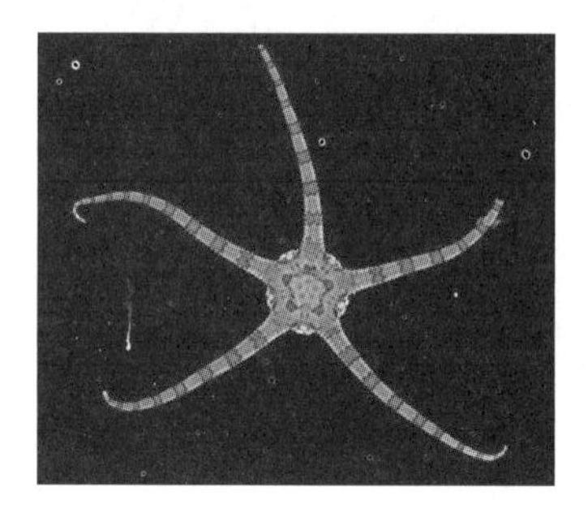

Fig. 10-3 Brittle stars

10–4 Class 3: Echinoidea

Their characteristics include: body is disc-like hemispherical (Fig. 10-4); they are devoid of arms or free-rays; they possess tube feet with a sucker; they possess compact skeleton and movable spines.

Examples: *Echinus*, *Cidaris*, *Arbacia*, *Echinocardium*, *Diadema* (Fig. 10-5).

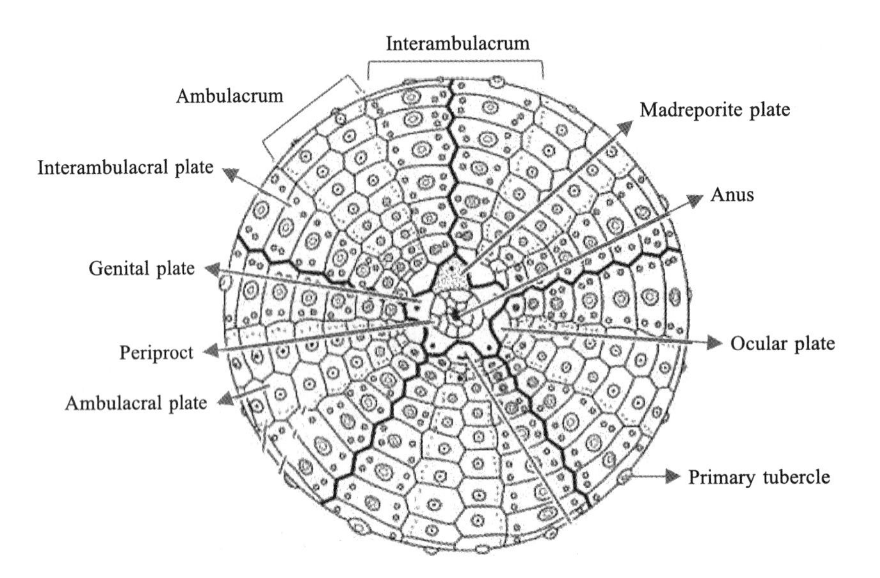

Fig. 10-4 Schematic of echinoids

Fig. 10-5 Echinoids

10-5 Class 4: Holothuroidea

Their characteristics include: body is elongated in the oral-aboral axis and it is like cucumber; they have no arms, spines and pedicellariae; the tube feet are sucking type which is modified into tentacles and form a circle around mouth; respiratory organ is cloacal respiratory tree (Fig. 10-6).

Examples: *Cucumaria*, *Holothuria*, *Mesothuria*, etc (Fig. 10-7).

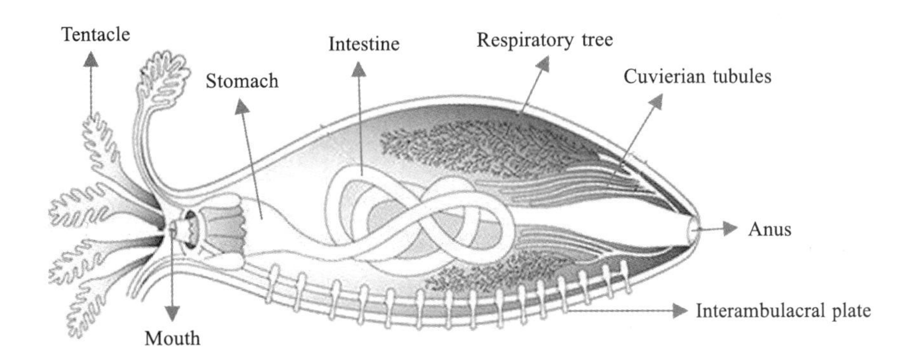

Fig. 10-6 Schematic of sea cucumbers

Fig. 10-7 Sea cucumbers

10–6 Class 5: Crinoidea

Their characteristics include: body is star shaped; some of the forms were extinct; arms bifurcated, with two pinnules; they have tube feet without suckers.

Examples: *Neometra*, *Antedon*, *Rhizocrinus*, etc (Fig. 10-8).

Fig. 10-8 Crinoids

10–7 学习要点

● 本门代表动物——海星的形态结构特点；棘皮动物是高等的后口动物，具真体腔和石灰质的内骨骼；棘皮动物幼体为两侧对称、成体为次生性辐射对称的适应意义。

● 棘皮动物特有的水管系统和围血系统。

● 棘皮动物的基本形态结构特点（体盘与棘、口面与反口面、步带和间步带）；海星的个体发育特点和再生能力。

● 棘皮动物门的分纲（海胆纲、海参纲、海星纲、蛇尾纲和海百合纲）。

● 海胆纲的主要特征；海胆的骨板结构（步带与间步带骨板，顶系的结构及骨板排列方式）。

● 海胆咀嚼器 Aristotle's lantern 的构造和功能；海胆的食性及其与海藻养殖业的关系。

● 海参纲的主要特征［蠕虫状体形；体具二道区和三道区；触手的形态多样（盾状触手、羽状触手、枝状触手和指状触手］；体内的呼吸树与居维氏器 Cuvierian organ 的结构；排脏现象的适应意义。

10-8 巩固测验

【名词】

原口动物、后口动物、呼吸树、水管系统、亚里士多德提灯

【选择题】

1. 以肠体腔法形成真体腔的动物是（　　）。

A. 环节动物　B. 软体动物　C. 节肢动物　D. 棘皮动物

2. 棘皮动物的成虫是辐射对称，其幼虫是（　　）。

A. 辐射对称　B. 两侧对称　C. 两侧辐射对称　D. 无对称

3. 不同类群的动物均有一些独有的特征。下列结构中，属于棘皮动物特有的是（　　）。

A. 后口　B. 水管系　C. 水沟系　D. 中胚层来源的骨骼

4. 中胚层的出现在动物的演化中具有重要的意义。下列结构中属于中胚层来源的是（　　）。

A. 红珊瑚的骨骼　B. 涡虫的肠壁　C. 乌贼的骨骼　D. 海参的骨针

5. 下列结构中不属于海星水管系统的是（　　）。

A. 环水管　B. 间辐管　C. 石管　D. 波氏囊

6. 下列动物中口和肛门均位于口面的动物是（　　）。

A. 海燕　B. 海胆　C. 海百合　D. 蛇尾

【简答题】

1. 棘皮动物的基本特征有哪些？

2. 为什么说棘皮动物是无脊椎动物中的高等类群？

Chapter 11 Phylum Chordata

11-1 Characteristics of Phylum Chordata

- Notochord—It is a longitudinal rod that is made of cartilage and runs between the nerve cord and the digestive tract. Its main function is to support the nerve cord. In vertebrate animals, the vertebral column replaces the notochord.
- Dorsal nerve cord—This is a bundle of nerve fibres which connects the brain to the muscles and other organs.
- Pharyngeal slits—They are the openings which connect the mouth and the throat. These openings allow the entry of water through the mouth, without entering the digestive system.
- Post-anal tail—This is an extension of the body beyond the anus. In some chordates, the tail has skeletal muscles, which help in locomotion.
- It will surprise you but this phylum is a very diverse phylum, with about 43,000 species. Most of these organisms can be found in the subphylum Vertebrata. In the animal kingdom, this is considered as the third largest phylum. Phylum Chordata is again divided into three subphyla. They are: Urochordata, Cephalochordata, and Vertebrata.

This phylum is probably the most notable phylum, as all human beings and other mammals and birds that are known to you fall under this phylum. The most distinguishing character that all animals belonging to this phylum have is the presence of notochord (Fig. 11-1).

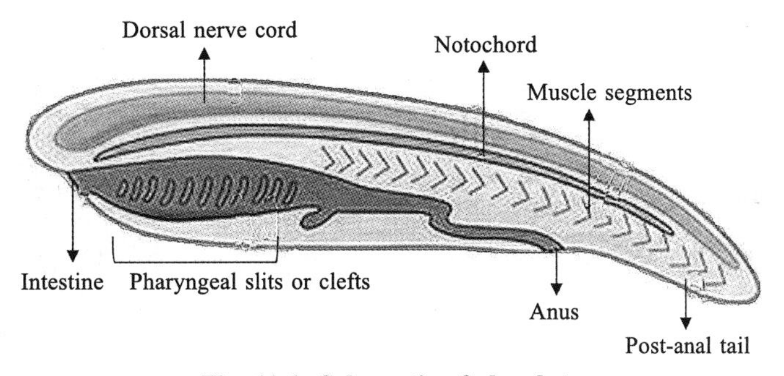

Fig. 11-1 Schematic of chordates

11-2 Subphylum Urochordata

Their characteristics include: the adults are fixed to the substratum; it is also known as Tunicata because the body of an adult is enclosed within a tunic made up of cellulose-like substance known as tunicin; notochord can be seen only in the larval stage and disappears in the adults; the nerve cord present in the larva is replaced by a dorsal ganglion in adults; the larva can move and undergoes metamorphosis (Fig. 11-2).

Examples: *Ascidia*, *Salpa*, *Doliolum*, *Botryllus* (Fig. 11-3).

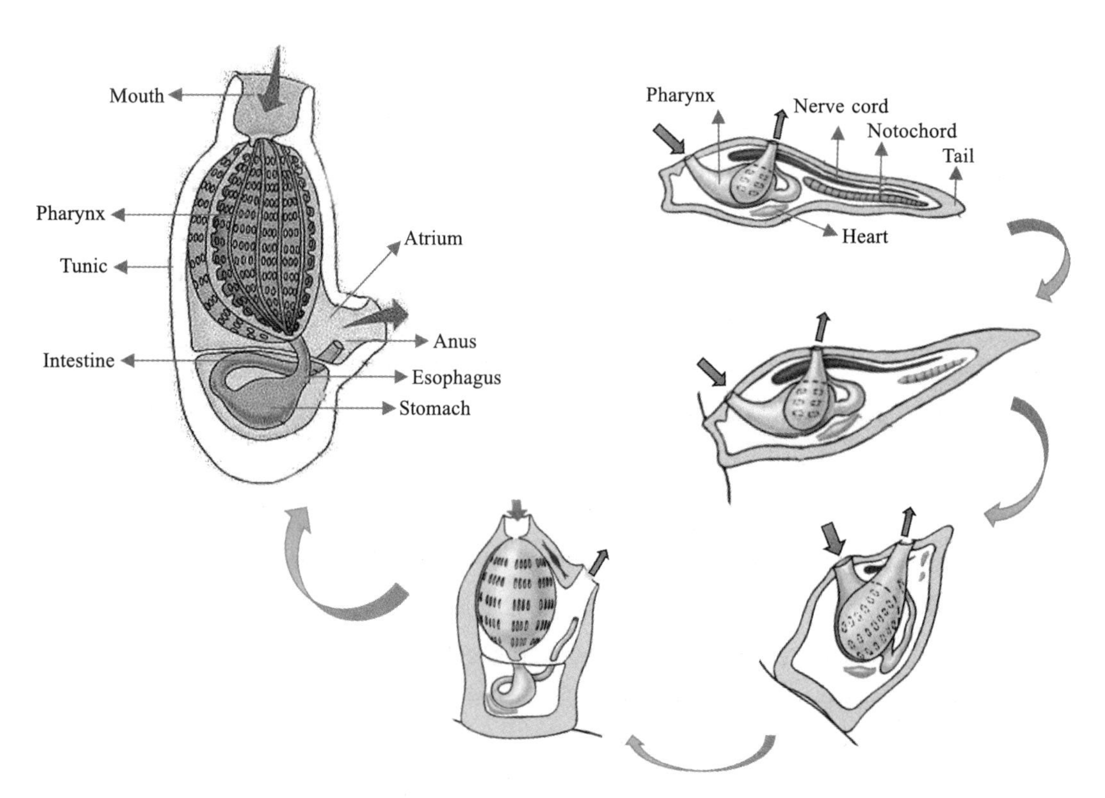

Fig. 11-2 Schematic and development of tunicates

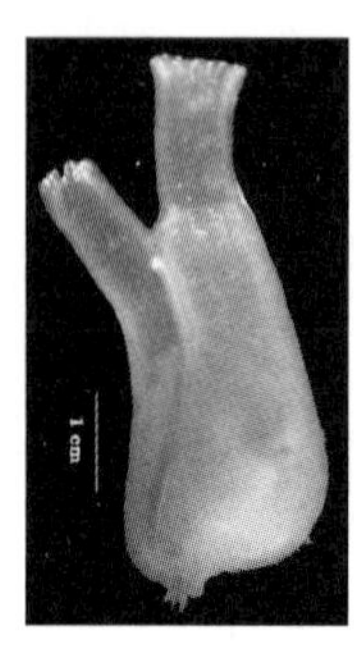

Fig. 11-3 Tunicates

11-3 Subphylum Cephalochordata

Their characteristics include: the atrium is present; the tail is present throughout life; they show progressive metamorphosis; the notochord is found throughout life; a large number of well-developed pharyngeal gill slits are present.

Lancelets (Fig. 11-4) possess the notochord and nerve cord throughout their life. However, they lack the brain and bony vertebral column, such as *Branchiostoma*.

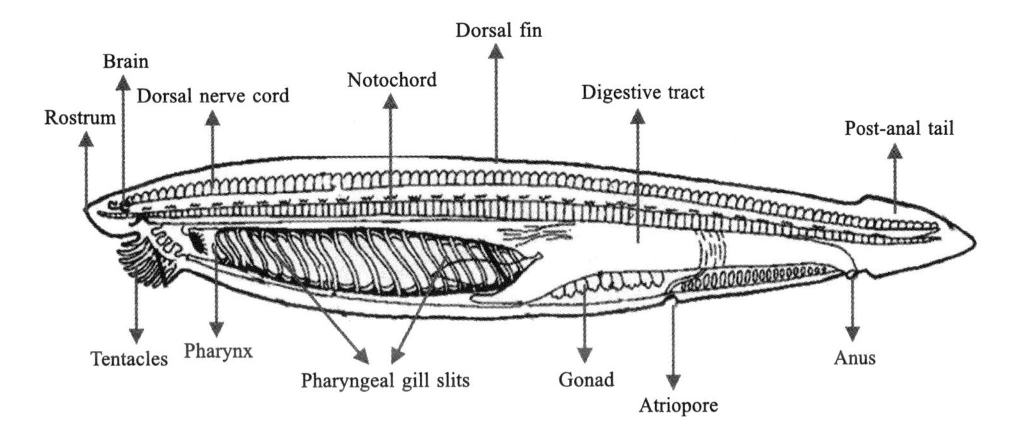

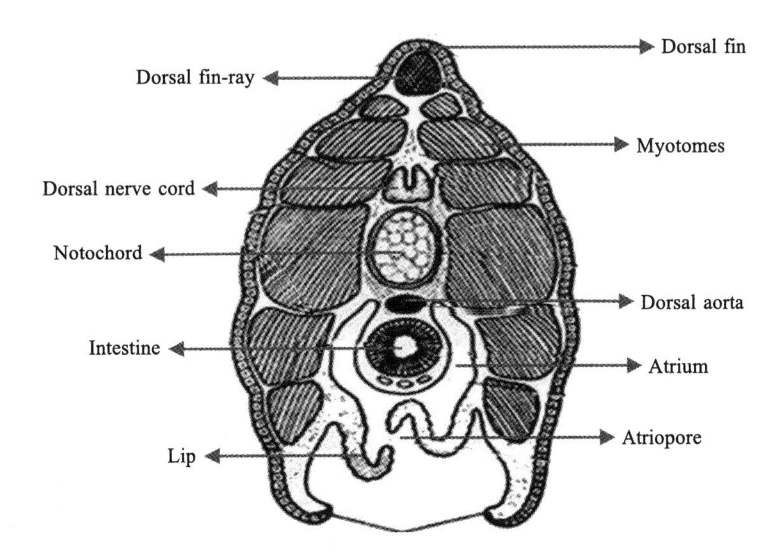

Fig. 11-4 Schematic of lancelets

11-4 Subphylum Vertebrata

Their characteristics include: these are advanced chordates and have cranium around the brain. The notochord is replaced by a vertebral column in adults. This is why it is said that "all vertebrates are chordates but all chordates are not vertebrates". A high degree of cephalization is observed. The epidermis is multi-layered. They consist of three types of muscles—striped, unstriped and cardiac. They have a well-developed coelom. The alimentary canal is complete. The heart is three or four-chambered. They have well-developed respiratory and excretory systems. Endocrine glands are present in all. They are unisexual and reproduce sexually, hagfish being an exception.

Subphylum Vertebrata is further classified into six classes. They are: Cyclostomata, Pisces, Amphibia, Reptilia, Aves, and Mammalia (Fig. 11-5).

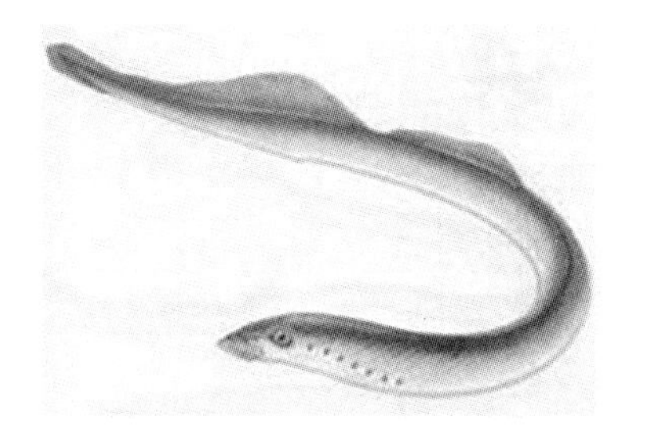

Fig. 11-5 Vertebrates

11-5 学习要点

● 脊索动物的主要特征和次要特征：①脊索 notochord；②背神经管 dorsal tubular nerve cord；③咽鳃裂 pharyngeal gill slits；④具后口、肛后尾；⑤闭管式循环系统；⑥心脏与主动脉位于消化道腹面；⑦具中胚层形成的内骨骼、分节的肌节。

● 脊索动物的分类；无头类、有头类、颌口类、无羊膜类、羊膜类等概念；尾索动物的主要特征及分纲；文昌鱼的主要特征；文昌鱼身体构造的两重性及在动物进化史上的意义；头索动物在动物系统发生中的作用和意义；尾索动物的逆行变态。

11-6 巩固测验

【名词】

逆行变态、脊索、变温动物、恒温动物

【选择】

1. 下列属于尾索动物亚门的动物是（　　）。

A. 文昌鱼　　B. 海鞘　　C. 鲤鱼　　D. 柱头虫

2. 下列动物成体中无脊索的是（　　）。

A. 海鞘　　B. 文昌鱼　　C. 七鳃鳗　　D. 盲鳗

3. 文昌鱼在科学研究上的重要意义体现在（　　）。

A. 食用　　B. 动物进化　　C. 培养新品种　　D. 开发动物智力

4. 成体具有脊索的动物是（　　）。

A. 文昌鱼　　B. 蛇蜥　　C. 青蛙　　D. 蜥蜴

5. 下列属头索动物亚门的动物是（　　）。

A. 文昌鱼　　B. 海鞘　　C. 黄鳝　　D. 大鲵

【简答】

1. 脊索动物的三大主要特征是什么？脊索动物还有哪些次要特征？
2. 脊索动物门分为几个亚门？脊椎动物亚门分为几个纲？
3. 为什么说尾索动物属于脊索动物？尾索动物有哪些主要特征？
4. 以文昌鱼为例，简述头索动物的主要特征。

Chapter 12 Class Cyclostomata

12-1 Characteristics of Cyclostomata

- The body is round and elongated like an eel.
- The paired fins are absent.
- Median fins have cartilaginous fin rays.
- Paired appendages are absent.
- The skin is soft and smooth, devoid of any scales.
- Spleen is absent.
- The exoskeleton is absent. The endoskeleton is cartilaginous with no bones.
- The notochord is present throughout their lives.
- The digestive system is devoid of any stomach.
- The nostril is single and median.
- The gills are 5–16 in pairs.
- The heart is two-chambered.
- The brain is visible.
- The lateral line acts as a sense organ.
- About 10 pairs of cranial nerves are present.
- The sexes are separate. Some hagfish species are believed to be hermaphrodite.
- A pair of mesonephric kidneys make up the excretory system.
- Development may be direct (*Myxine*) or indirect (*Petromyzon*).
- The cyclostomes are subdivided into two major orders.

Cyclostomata is a group of agnathans, which comprises the living jawless fishes: the lampreys and hagfishes. Both groups have jawless mouths with horny epidermal structures that function as teeth, and branchial arches that are internally positioned instead of external as jawed fishes (Fig. 12-1). They are parasitic, usually feeding on fish in their adult stage (Fig. 12-2). Morphologically, they resemble eels. They are known to be the only living vertebrates without true jaws, hence called Agnatha.

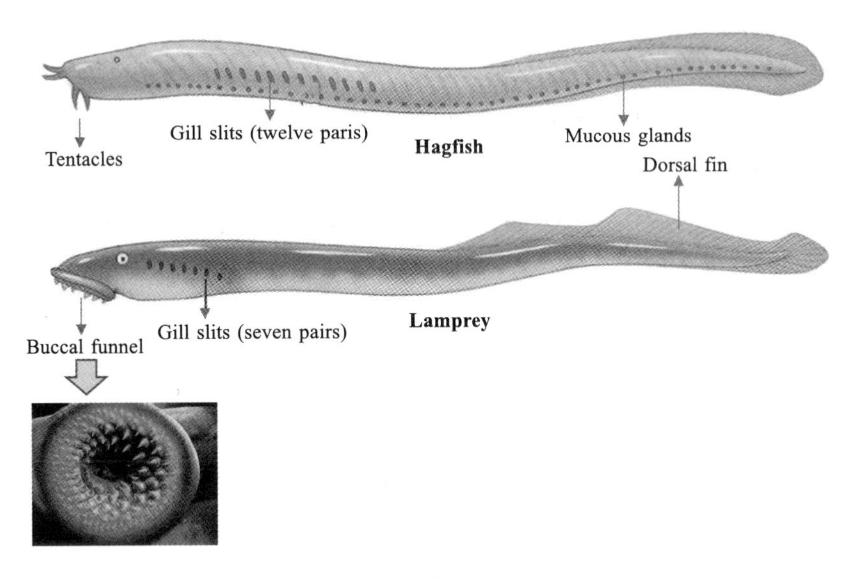

Fig. 12-1 Schematic of lampreys and hagfishes

Fig. 12-2 Cyclostomes feeding on fish

12–2 Order Petromyzontiformes

Lampreys or lamper eels belong to this order. Their characteristics include: they are found in both marine and freshwater; they have a ventral mouth with many horny teeth; the nostril is present dorsally; they possess a well-developed dorsal fin; the dorsal and ventral roots of spinal nerves are separate; the development is indirect.

Examples: *Petromyzon*, *Lampetra*.

12-3 Order Myxiniformes

Hagfishes represent this order. Their characteristics include: they are found exclusively in the marine environment; they have a terminal mouth with few teeth; they have no buccal cavity; the nostril is terminal; they possess 6-14 pairs of gill slits; the dorsal and ventral roots of the spinal nerves are not separate; eggs are large and few in number; the dorsal fin is usually absent, or weak.

Examples: *Myxine*, *Eptatretus*.

12-4 学习要点

- 圆口纲的原始特征和一般特征；七鳃鳗*Petromyzon*和盲鳗*Myxine*的主要区别。
- 无颌类的特征。

12-5 巩固测验

【名词】

鳃囊、口漏斗

【简答】

为什么说圆口纲在脊椎动物类群中比较低等？

Chapter 13 Class Pisces

13–1 Characteristics of Pisces

- Aquatic, either freshwater or marine, herbivorous or carnivorous, cold blooded, oviparous or ovoviviparous vertebrates.
- Body usually streamlined, spindle-shaped, some are elongated snake-like and a few are dorsoventrally compressed, and differentiated into head, trunk and tail.
- Locomotion by paired pectoral and pelvic fins along with median dorsal and caudal fins, supported by true dermal fin-rays. Muscular tail used in propulsion.
- Exoskeleton of dermal scales, denticles or bony plates (in Placodermi) covering body surface. Placoid in Chondrichthyes and ganoid, cycloid or ctenoid in Osteichthyes.
- Endoskeleton is cartilaginous or bony. The notochord is usually replaced by vertebrae, either bone or cartilage. Presence of well-developed skull and a system of visceral arches, of which the first pair forms the upper and lower jaws, the latter movably articulated with the skull.
- Muscles arranged into segments called myotomes, with separate dorsal and ventral parts.
- Alimentary canal with definite stomach and pancreas and terminates into cloaca or anus.
- Organs of respiration are gills. Gill-slits 5–7 pairs, naked or covered by an operculum.
- Heart is venous and two chambered, i.e. one auricle and one ventricle. Sinus venosus and renal and portal systems present. Erythrocytes nucleated. Poikilothermous.
- Kidneys mesonephros. Excretions ureotelic.
- Brain with usual five parts. Cranial nerves ten pairs.
- Nostrils are paired but do not open into pharynx except dipnoi. Nasal capsules are partly separate in Chondrichthyes and completely separate in Osteichthyes.
- Tympanic cavity and ear ossicles are absent.
- Internal ear with three semicircular canals.
- Lateral line system is well developed.
- Sexes separate. Gonads typically paired. Gonoducts open into cloaca or independently.

- Fertilization internal or external. Females of Chondrichthyes are oviparous or ovoviviparous and of Osteichthyes are mostly oviparous and rarely ovoviviparous or viviparous. Eggs with large amount of yolk. Cleavage meroblastic.
- Extra-embryonic membranes are absent.
- Development usually direct without or with little metamorphosis.

The Superclass Pisces are the truly jawed vertebrates. They have organs of respiration and locomotion related to a permanently aquatic life. The respiratory organs are the gills and the organs of locomotion are paired and unpaired fins. All are poikilothermous.

- **External Features**

The body is spindle-shaped, higher than wide, and of oval cross section for easy passage through the water. The head extends from the tip of the snout to the hind edge of the operculum, the trunk from this point to the anus, and the remainder is the tail. The large mouth is terminal, with distinct jaws that bear fine teeth. Dorsally on the snout are two double nostrils (olfactory sacs), the eyes are lateral without lids, and behind each is a thin bony gill cover, or operculum, with free edges below and posteriorly. Under each operculum are four comblike gills. The anus and urogenital aperture precede the anal fin (Fig. 13-1). Ichthyologists measure the standard length of a fish from the tip of the snout to the end of the last vertebra (hypural plate) to avoid error from wear of the tail fin, but fishermen include the fin.

- **Body Covering**

The entire fish is covered by soft mucus-producing epidermis that protects against abrasion and entry of disease organisms. The trunk and tail bear thin rounded dermal scales, in lengthwise and diagonal rows, their free posterior edges overlapping like shingles on a roof; each lies in a dermal

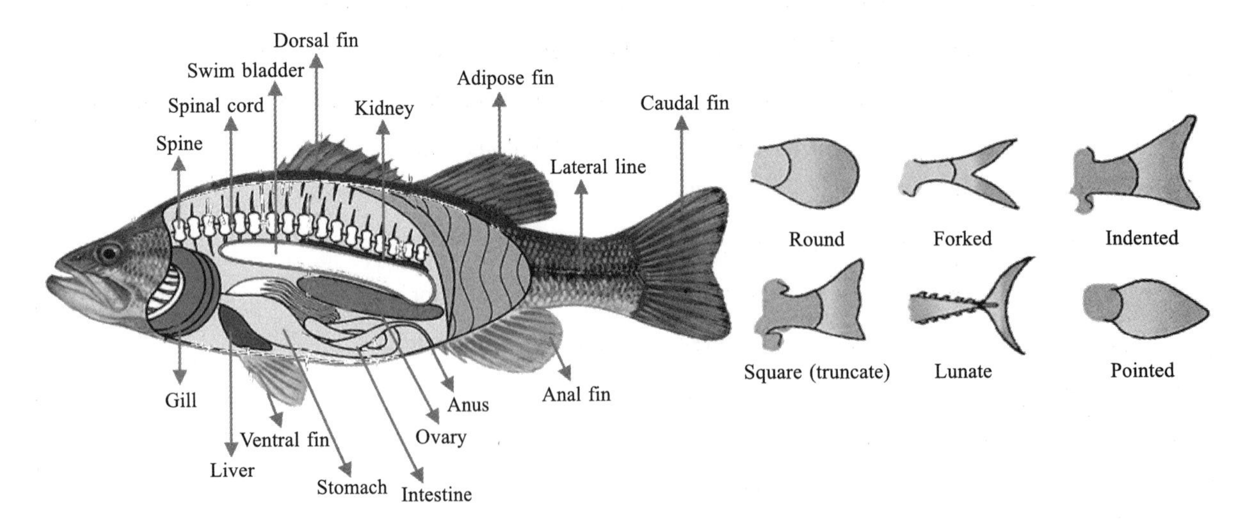

Fig. 13-1 Schematic of fish and types of caudal fin

pocket and grows throughout life. The free portion is covered with a thin layer of skin. The lateral line, along either side of the body, is a row of small pores connected to a lengthwise tubular canal under the scales.

● **Skeleton**

The scales and fins constitute an exoskeleton. The endoskeleton consists of the skull, vertebral column, ribs, pectoral girdle, and many small accessory bones (pterygiophores) supporting the fin rays. The skull comprises the cranium housing the brain, the capsules for the paired organs of special sense (olfactory, optic, auditory), and the visceral skeleton, which provides the jaws and the supports for the tongue and gill mechanism. The skull has a double articulation with the vertebral column and is so firmly joined that a fish cannot "turn its head". Teeth are usually present on the premaxillary, dentary, vomer, and palatine bones.

● **Scales**

Most bony fishes are covered with scales (Fig. 13-2). While usually thin and overlapping, the scales are separate and minute on eels, small and tubercle-like on some flounders, and slender spines on porcupinefishes. The overlapping scales of tarpon are up to 5 cm in width; some fishes are scaleless or naked (some catfish). The scales are usually bony. On the perch and many others, the exposed hind part of each has many tiny spines, making a ctenoid scale. Others that lack such spines are termed cycloid scales, and still others have ganoid scales, which are bony and capped with ganoin, a hard glassy, enamel-like substance. Ganoid scales were characteristic of the early ray-finned fishes (palaeoniscids and holosteans) but in living teleosts are present only in the bichirs (*Polypterus*), reedfish (*Erpetoichthys*), and garpikes (*Lepisosteus*). The heads and bodies of various fishes, living and fossil, are armor-plated with large stout scales, as in the trunkfish; living species so protected are usually small or sluggish.

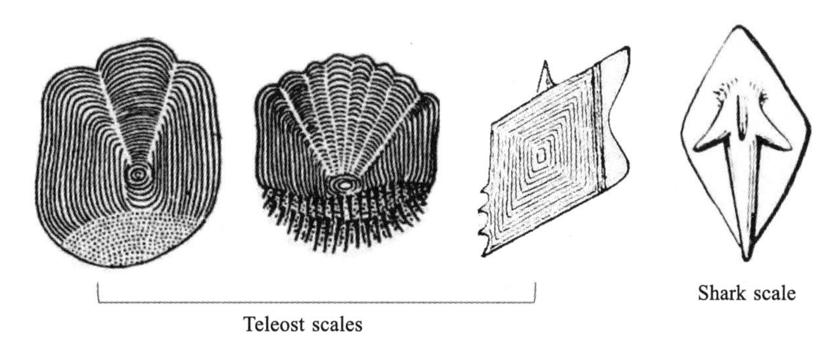

Fig. 13-2 Scales

● **Muscular System**

The substance of the trunk and tail consists chiefly of segmental muscles (myomeres) that alternate with the vertebrae and produce the swimming and turning movements. Fish myomeres are broadly Σ-shaped, in four principal bands, and heaviest along the back. Between successive myomeres are delicate connective tissue septa; when a fish is cooked, these dissolve to leave the myomeres as individual "flakes". The muscles of the fins, gill region, and head are small.

● **Digestive System**

The jaws have many small conical teeth to grasp food, and farther back are pharyngeal and gill raker teeth helpful inholding and crushing it. Mucous glands are numerous, but there are no salivary glands. The small tongue is attached to the floor of the mouth cavity and may aid in respiratory movements. The pharynx has gills on the sides and leads to a short esophagus followed by the recurved stomach. A pyloric valve separates the latter from the intestine. Three tubular pyloric caeca, secretory or absorptive in function, attach to the intestine. There is a large liver anteriorly in the body cavity, a gall bladder, and a bile duct to the intestine. The pancreas is usually diffuse (Fig. 13-3).

● **Circulatory System**

The two-chambered heart lies below the pharynx in the pericardial cavity, an anterior portion of the coelom. Venous blood passes into the sinus venosus, to the thin-walled atrium, thence into the muscular ventricle, all separated by valves that prevent reverse flow (Fig. 13-4). Rhythmic contractions of the ventricle force the blood through the conus arteriosus and short ventral aorta into four pairs of afferent branchial arteries distributing to capillaries in the gill filaments for oxygenation. It then collects in correspondingly paired efferent branchial arteries leading to the dorsal aorta, which distributes branches to all parts of the head and body. The principal veins are the paired anterior cardinals and posterior cardinals and the unpaired hepatic portal circulation leading through the liver. The blood of fishes is pale and scanty as compared with that of terrestrial vertebrates. The fluid plasma contains nucleated oval red cells (erythrocytes) and various types of white cells (leukocytes). The spleen, a part of the blood system, is a large red-colored gland near the stomach. A lymphatic system is also present.

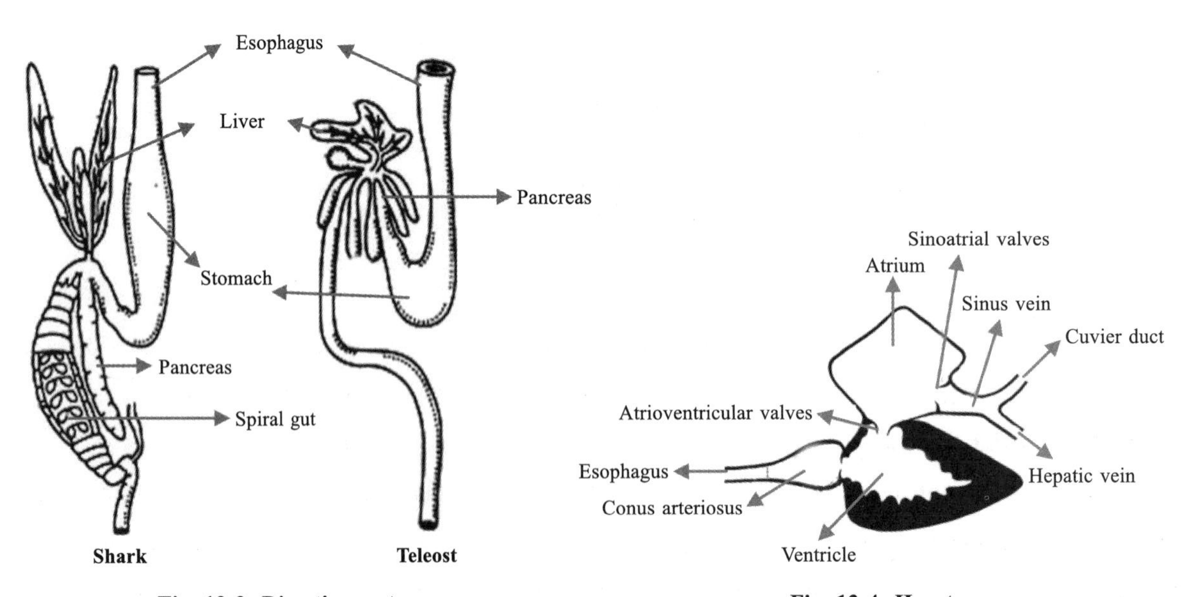

Fig. 13-3 Digestive system

Fig. 13-4 Heart

● **Respiratory System**

The perch respires by means of gills, of which there are four in a common gill chamber on each side of the pharynx, beneath the operculum. A gill consists of a double row of slender gill filaments; every filament bears many minute transverse plates covered with thin epithelium and containing capillaries between the afferent and efferent branchial arteries. Each gill is supported on a cartilaginous gill arch, and its inner border has expanded gill rakers, which protect against hard particles and keep food from passing out the gill slits.

● **Excretory System**

The two slender dark kidneys lie dorsally between the swim bladder and vertebrae. Fluid nitrogenous waste (ammonia, urea) removed from the blood is carried posteriorly from each in a tubular ureter, both emptying into a urinary bladder, which in turn discharges through the urogenital sinus to the exterior. Diffusion outward of these substances also occurs across the gills.

● **Nervous System**

The perch brain is short. The olfactory lobes, cerebral hemispheres, and diencephalon are smaller than those in a shark and the optic lobes and cerebellum are larger. There are 10 pairs of cranial nerves. The nerve cord is covered by the neural arches and gives off a pair of lateral spinal nerves to each body segment.

● **Organs of Special Sense**

The lateral line system, also called lateralis system, a system of tactile sense organs, unique to aquatic vertebrates from cyclostome fishes (lampreys and hagfish) to amphibians, serves to detect movements and pressure changes in the surrounding water.

● **Reproductive System**

In a male, the two testes enlarge greatly in the breeding season, and at mating the "milt", or sperm, passes in a ductus deferens from each to emerge from the urogenital aperture. In a female, the eggs pass from the two united ovaries through the oviducts.

13-2 Classifications of Pisces

About 40,000 species of fishes are known. Various researchers have proposed different schemes of classification of fishes. However, no classification has been universally accepted because of confusion due to large number of fishes and great diversity in their shape, size, habit and habitat.

Romer (1959) classified the fishes into two classes—Chondrichthyes (includes all cartilaginous fishes) and Osteichthyes (includes all bony fishes like Dipnoi and Teleostomi). Osteichthyes is divided into two subclasses—Sarcopterygii and Actinopterygii. Parker and Haswell (1962) have further combined all the extinct jawed fishes under a single class—Placodermi or Aphetohyoidea.

● **Cartilaginous Fishes**

Examples of cartilaginous fishes are shown in Fig. 13-5.

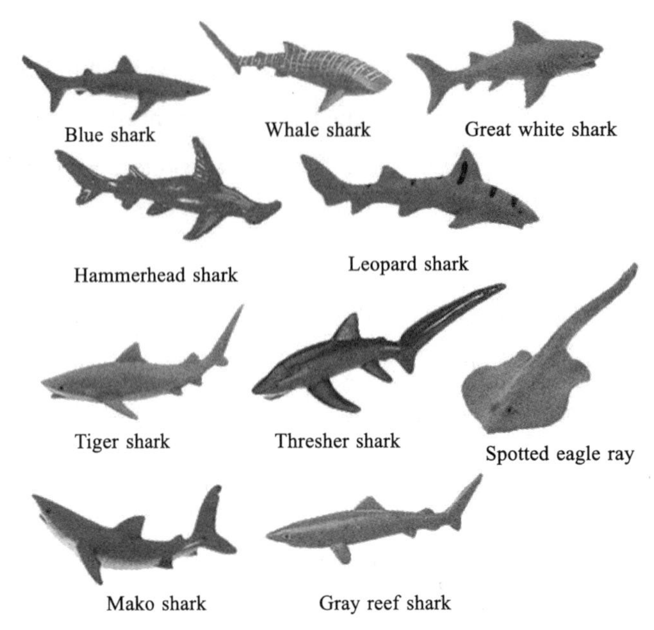

Fig. 13-5 Cartilaginous fishes

● **Osteichthyes**

Examples of Osteichthyes are shown in Fig. 13-6 and Fig. 13-7.

(1) Teleostomi

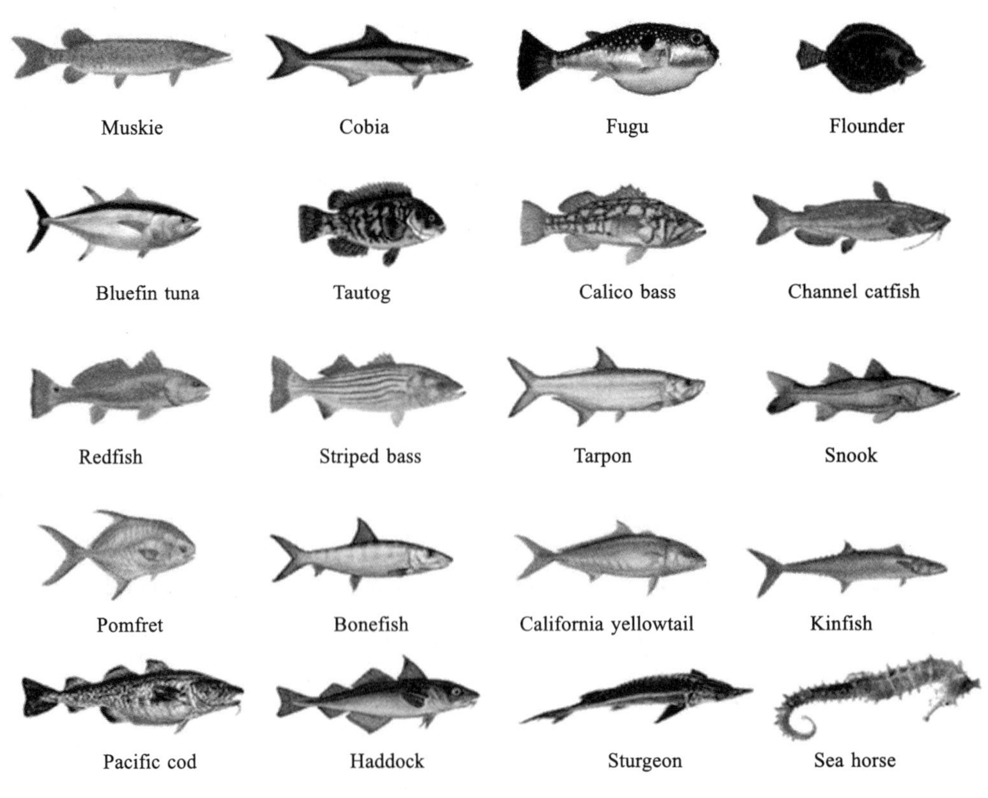

Fig. 13-6 Teleostomi

(2) Dipnoi

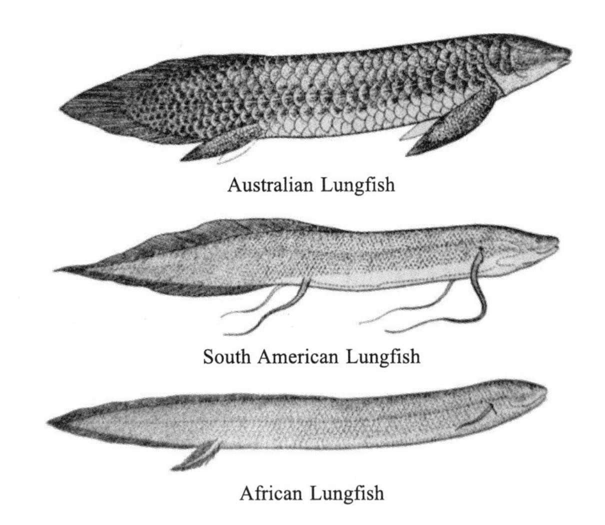

Fig. 13-7 Dipnoi

13-3 学习要点

● 鱼纲的主要特征（鱼的基本体形，口位，鳞的类型，脑颅、咽颅及它们的连接方式，初生颌和次生颌，动脉弓的构成，等等）。

● 鱼类适应水生生活的特征。

● 鳃呼吸与单循环。

13-4 巩固测验

【名词】

洄游、测线、韦伯氏器、螺旋瓣、动脉圆锥

【选择】

1. 圆口类动物终身保留（　　）。

A. 脊椎　　B. 脊柱　　C. 脊索　　D. 脊髓

2. 脊椎动物中种类最多的类群是（　　）。

A. 鱼纲　　B. 爬行纲　　C. 鸟纲　　D. 哺乳纲

3. 鱼的身体分为（　　）。

A. 头、躯干、附肢　　B. 头、胸、腹、尾

C. 头、躯干、尾　　D. 头、躯干、尾、四肢

4. 硬骨鱼脊椎的分区是（　　）。

A. 颈椎、躯干椎、尾椎　　B. 胸椎、腰椎、尾椎

C. 躯干椎、尾椎　　D. 颈椎、躯干

5. 软骨鱼肠管内具有（　　）。

A. 幽门盲囊　　B. 螺旋瓣

C. 脂肪体　　D. 黏膜褶

6. 下列属于生殖洄游的是（　　）。

A. 越冬场—育肥场　　B. 育肥场—越冬场

C. 产卵场—育肥场　　D. 越冬场—产卵场—育肥场—产卵场

7. 在鱼类中，雄鱼具有育儿囊的是（　　）。

A. 海马　　B. 鲨鱼

C. 海鳗　　D. 海鳝

8. “四大家鱼”指的是（　　）。

A. 鲤、鲫、鲢、鳙　　B. 鲤、鲫、鳢、鲂

C. 带鱼、鲳鱼、鲈鱼　　D. 青、草、鲢、鳙

【简答】

1. 软骨鱼与硬骨鱼有何差异？

2. 鱼类的主要类群是如何进行渗透压调节，保持渗透压平衡的？

3. 鱼鳞的类型有哪些？鱼鳍的类型有哪些？

Chapter 14 Class Amphibia

14–1 Characteristics of Amphibians

- They are poikilothermic animals and most forms hibernate in winter and aestivate in summer.
- The body may be long and narrow, short and broad, depressed or cylindrical. The body may be divisible into head, trunk and tail (salamander) or only into head and trunk (frog) (Fig. 14-1).
- There are two pairs of pentadactyl limbs, each with 4–5 or fewer digits. The digits are without claws, nails or hooves, and often with webs.
- The skin is without scales, smooth, moist, rich in multicellular mucous or poison glands.
- Skull is flat and dicondylic (with two occipital condyles).
- The mouth is large and armed with acrodont and vomerine teeth in the upper jaw (Fig. 14-2). They have a true tongue and it is soft, mucous-coated and attached at the front end (protrusible tongue). Distinct liver and pancreas are present. Alimentary canal leads into the cloaca.
- Tadpoles respire through gills as they live in water. Adults respire with lungs, lining of buccopharyngeal cavity and skin. Some forms have vocal cords.
- Circulatory system is closed and the heart is three chambered: two auricles and one ventricle. The ventricle pumps the mixed blood; sinus venosus and conus arteriosus are present. Well-developed renal portal system and hepatic portal system are present. Red blood cells are oval, biconvex and nucleated.
- Kidneys are pronephric in larvae and mesonephric in adults. Waste material is either urea (in ureotelic or tailless forms) or ammonia (in ammonotelic forms or larvae and tailed forms). Urinary bladder is present.
- The gonoducts lead into the cloaca. Males lack a copulatory organ. Fertilization may either be external or internal. Most forms are oviparous; eggs have a coat of jelly and usually laid in water.

- The olfactory sacs are paired and dorsal. Eyes often have movable lids. Middle ear with a single auditory ossicle is present in addition to the internal ear. Tympanum covers the middle ear but there is no external ear (Fig. 14-3). Lateral-line sense organs are present in the larvae and in aquatic forms but absent in terrestrial forms. There are 10 pairs of cranial nerves and 9–10 pairs of spinal nerves.

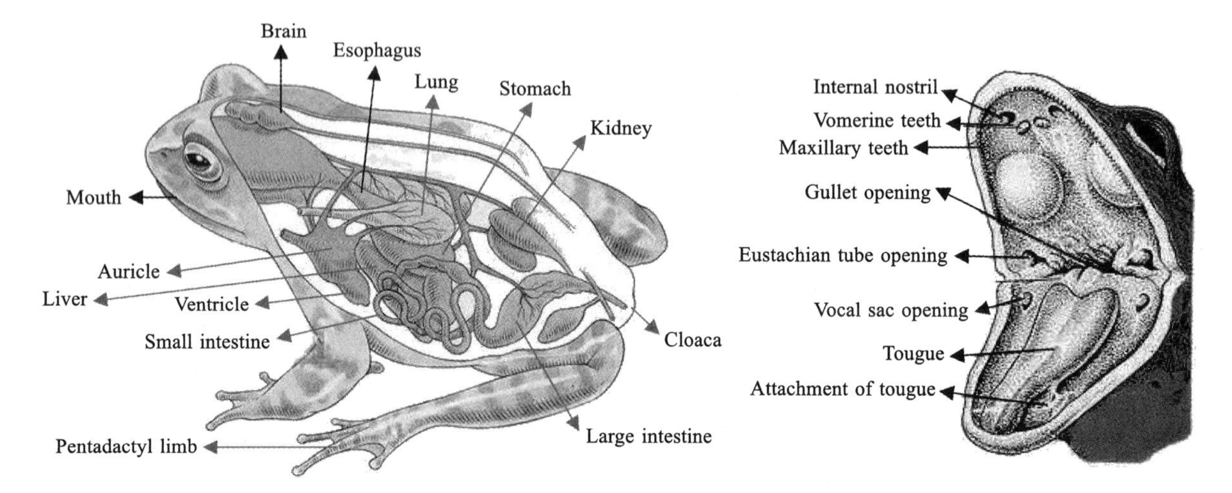

Fig. 14-1 Schematic of frogs

Fig. 14-2 Mouth of frogs

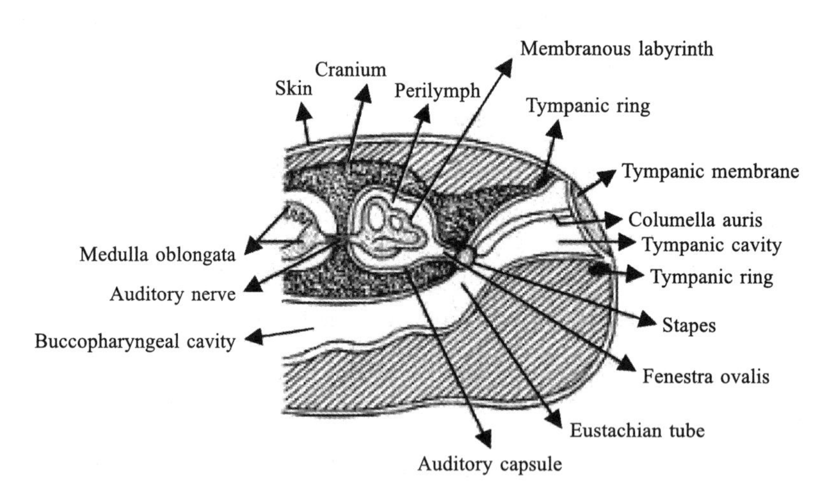

Fig. 14-3 Auditory system

Amphibians are animals that are characterized by their ability to survive both in water and on land. The name "amphibian" is derived from the Greek word "amphibious" which means "to live a double life". There are over 6,500 living species of amphibians with the majority of the species living within fresh aquatic water ecosystem. Most of them are born in water and start off as a larva and develop a land-based lifestyle as they develop. 90% of all amphibian species are frogs.

The three living orders of amphibians vary greatly in size and structure. The presence of a long tail and two pairs of limbs of about equal size distinguishes newts and salamanders (order Caudata) from other amphibians, although members of the eel-like family Sirenidae have no hind limbs.

Amphibians occur widely throughout the world, even edging north of the Arctic circle in Eurasia; they are absent only in Antarctica, most remote oceanic islands, and extremely xeric (dry) deserts. Frogs and toads show the greatest diversity in humid tropical environments. Salamanders primarily inhabit the Northern Hemisphere and are most abundant in cool, moist, montane forests; however, members of the family Plethodontidae, the lungless salamanders, are diverse in the humid tropical montane forests of Mexico, Central America, and northwestern South America. Caecilians are found spottily throughout the African, American, and Asian wet tropics.

Amphibians, especially anurans, are economically useful in reducing the number of insects that destroy crops or transmit diseases. Frogs are exploited as food, both for local consumption and commercially for export, with thousands of tons of frog legs harvested annually. The skin secretions of various tropical anurans are known to have hallucinogenic effects and effects on the central nervous and respiratory systems in humans. Some secretions have been found to contain magainin, a substance that provides a natural antibiotic effect. Other skin secretions, especially toxins, have potential use as anesthetics and painkillers. Biochemists are currently investigating these substances for medicinal use.

● **External features**

The highly vascular skin with its many mucous glands keeps the body surface moist in air and makes cutaneous respiration possible. The latter feature has restricted evolution of terrestrial adaptations in these animals. The skin contains poison glands, sometimes in special warts on toads, dorsolateral folds on frogs, and upper surface of tail on some salamanders. The skin of salamanders and caecilians adheres closely to the body muscles, but is more loosely attached in frogs and toads, which have extensive subdermal lymph spaces. The skin may be smooth and glistening or variously roughened with tubercles or warts. The coloration of amphibians often is bright and contrasted-green, yellow, red, brown, or black. The basic pattern is fixed on most species, but others undergo great changes in coloration under different environmental conditions.

In the leopard frog and others, three kinds of chromatophores form a system responsible for changes in color. Just beneath the epidermis are xanthophores; alone they impart a yellowish hue. Next below are the iridophores with light-reflecting organelles that produce light scattering Tyndall blue—a structural color. This filtered by the overlying yellow pigment results in the green skin color. The black melanophores are extensively branched cells under the iridophores and surrounding them with fingerlike processes that extend between them and the xanthophores. The skin darkens when melanin granules move outward from the cell body into the branches and obscure the iridophores. The pituitary hormone intermedin causes dispersal of melanin and con-traction of the iridophores. A reverse movement of both produces a paler skin. Various species of amphibians differ in details of pigmentation. Dark pigment in the epidermis usually is scant in those that change color, since a heavy layer would conceal the underlying pigments.

Thc palms and solcs of thc fcet bear small cornified tubercles; on the hind foot in the spadefoots (*Scaphiopus*) and some other toads, the innermost tubercle becomes a horny cutting spade that these animals use to dig backward into the ground when seeking shelter. The treefrogs (Hylidae), some other frogs (Dendrobatidae, Rhacophoridae), and some *Rana* in Asia have expanded disks on all toes by which they can adhere to and climb on vertical surfaces. Webbing of the hind toes is least in strictly terrestrial species but extensive in the more aquatic forms.

● **Skeleton and muscles**

The skeleton in salamanders resembles that of other elongate vertebrates in having many vertebrae (up to 100 in Amphiuma); some caecilians have over 250. Ribs are present in salamanders, caecilians, and some primitive frogs (Fig. 14-4). The pelvic girdle of salamanders is short, and caecilians lack both a pectoral and pelvic girdle. Most amphibians have fine teeth on the upper jaw and roof of the mouth; some salamanders have teeth on both jaws, but toads (Bufonidae) are toothless. The teeth are fastened to the surfaces of the bones and are continually and alternately replaced.

Segmental muscles are conspicuous on the trunk and tail of salamanders and trunk of caecilians. Gill-bearing species have special muscles to move the gills and to open or close the gill slits. The upper segment of each limb stands out laterally from the body, as in four-legged reptiles.

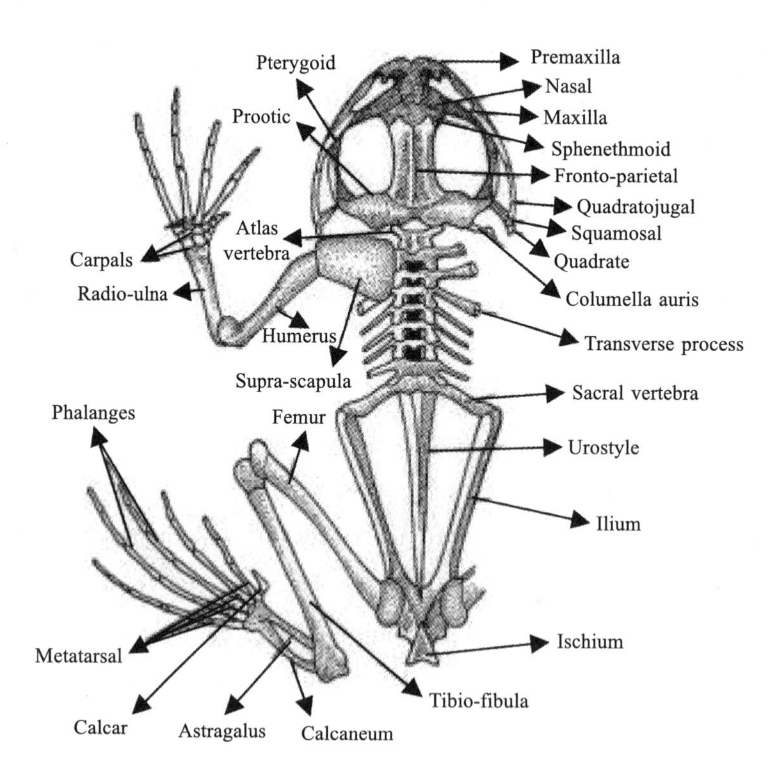

Fig. 14-4 Skeleton of frogs

● **Respiration**

Amphibians have more means for respiration than any other animal group, reflecting the transition from aquatic to land habitats. In different species, the gills, lungs, skin, and buccopharynx serve separately or in combination. The heart in larvae is fishlike, with one atrium and a ventricle. It receives only unoxygenated blood, which is pumped directly to the gills. Adults have two atria and one ventricle. The left atrium receives oxygenated blood from the skin and lungs, the right receives unoxygenated blood from the general circulation. Despite the single ventricle, there is some separation of atrial blood; that sent to respiratory surfaces is almost all unoxygenated (Fig. 14-5).

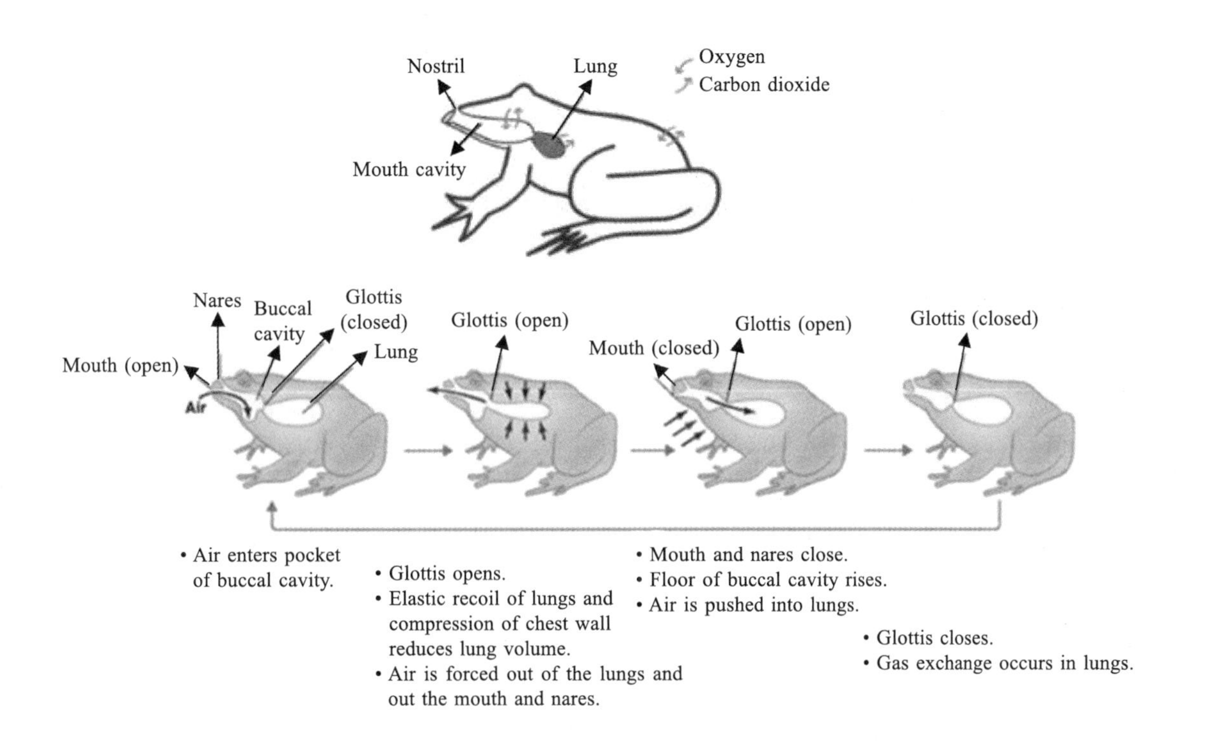

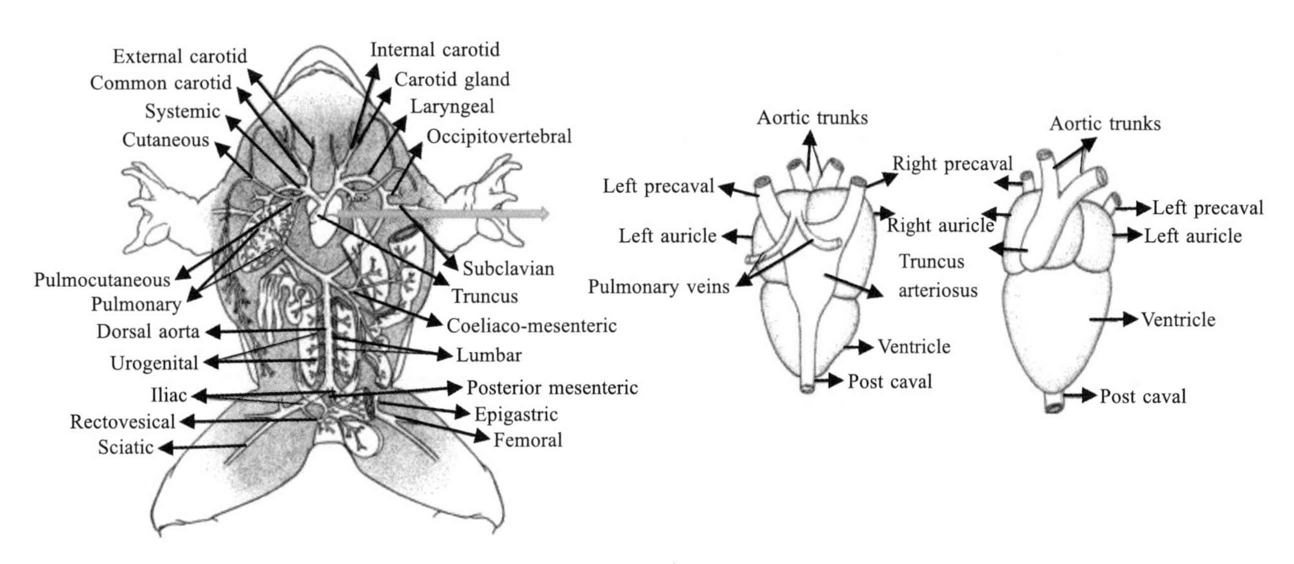

Fig. 14-5 Respiration and circulatory system of frogs

Three pairs of external gills occur in most embryos and larvae and persist in the adults of some strictly aquatic salamanders. In tadpoles, water is drawn in through the mouth and nostrils, then forced over the gills and out the spiracle(s). Salamanders aid aeration by moving their gills sometimes rhythmically. Amphibians have simple lungs, with low internal partitions containing blood vessels. In aquatic species, the lungs also serve as hydrostatic organs, being inflated when the animals are floating. In some salamanders that inhabit swift mountain streams (*Rhyacotriton*), the lungs are reduced, and all American land salamanders (Plethodontidae) lack lungs. The skin of all amphibians contains many blood vessels that aid aeration of the blood; this permits some aquatic species to remain submerged for long periods and to hibernate in ponds. Many species have buccopharyngeal respiration; pulsations of the throat move air in and out of the mouth cavity, and aeration of the blood occurs in vessels in the mucous membrane there. The proportion of capillaries, however, is small (0.5%–10% of those present in the skin).

Vocal cords in the larynx of frogs and toads serve to make the familiar calls, distinctive for each species, that serve to bring the sexes together for mating—chiefly in the spring. Some species have resonating pouches in the throat that amplify the sounds. Salamanders lack vocal cords, but a few kinds make faint squeaks. *Dicamptodon*, the one exception having vocal cords, makes a growling or barking sound.

● **Food**

Adult amphibians and the larvae of salamanders eat live moving animals, such as insects, worms, and small mollusks. Large aquatic species also take small fishes, the bullfrog sometimes catches small fishes, birds, or mammals, and big amphibians will devour small individuals of their own or other species. The aquatic larvae of toads and frogs feed mainly on algae and on bits of dead animals in the water. Some filter bacteria and other small organisms from the water. *Xenopus* tadpoles can remove particles only 1.1 μm in diameter.

● **Reproduction**

Most amphibians mate in water, where their eggs are deposited and hatch and where the resulting larvae live and grow until they metamorphose into the adult stages. Each species has a characteristic type of breeding place such as a large quiet lake or pond, a stream, or a transient pool; some breed on land. Male toads and frogs, upon entering the water, begin croaking to attract females. As each "ripe" female enters, she is clasped by a male, who clings on her back. As she extrudes her eggs, the clasping male discharges sperm or "milt" over them to effect fertilization. Many aquatic and terrestrial salamanders have courting performances in which the male noses about or may mount the female, but he eventually deposits one or more gelatinous spermatophores on the bottom of a stream or pond or on the ground. The spermatophore is a sperm packet on a gelatinous base. It is taken into the female's cloaca, where the spermatozoa are stored in the seminal receptacle, later to fertilize her eggs internally before they are laid.

● **Life cycle of frogs**

Life cycle of frogs is shown in Fig. 14-6.

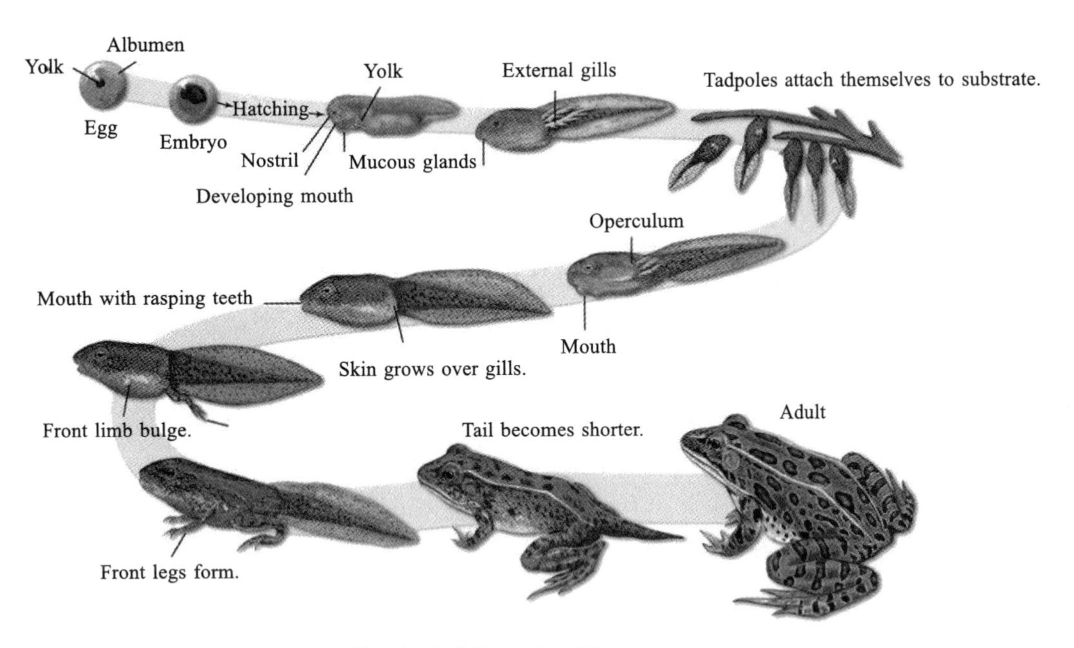

Fig. 14-6 Life cycle of frogs

14–2 Order 1: Apoda

Animals are limbless. Scales are present. They are called "blind worms" or caecilians.

Examples: *Uraeotyphlus*, *Ichthyophis*, *Siphonops* (Fig. 14-7).

Fig. 14-7 South American caecilian (*Siphonops annulatus*)

14–3 Order 2: Urodela

Tail is present.

Examples: *Necturus* (mud puppy), *Amphiuma* (Congo eel), *Salamandra*, *Proteus*, *Siren* (mud eel), *Ambystoma*, *Triturus* (newt), *Tylototrition* (crocodile newt) (Fig. 14-8).

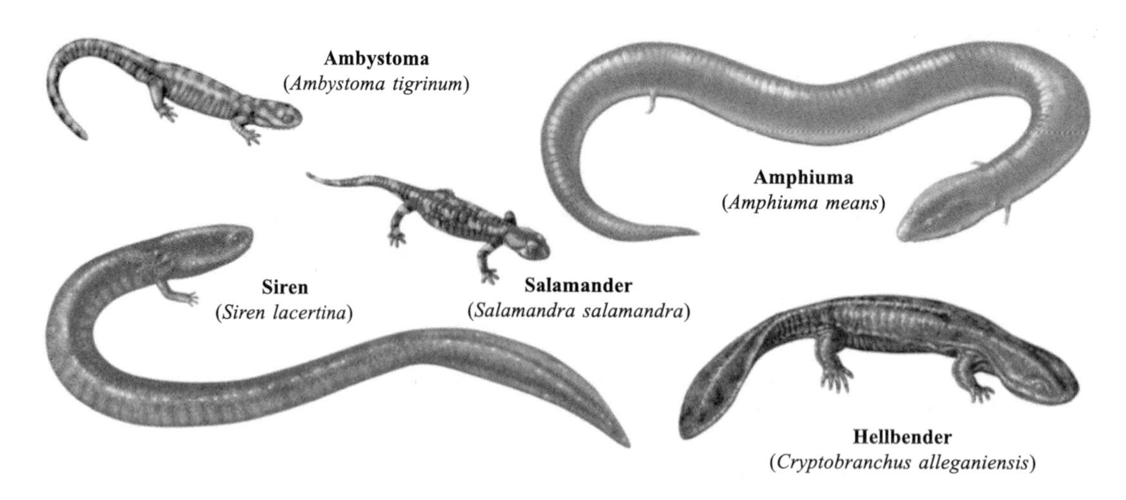

Fig. 14-8 Salamanders

14-4 Order 3: Anura

Tail is absent.

Examples: *Rana* (common frog), *Rhacophorus*, *Bufo*, *Hyla*, *Alytes*, *Xenopus* (African toad), *Pipa* (Fig. 14-9).

Fig. 14-9 Frogs

14-5 学习要点

● 五趾型附肢的特征及在脊椎动物进化史上的意义。

● 两栖类对陆地环境的初步适应及不完善性。

● 两栖纲的一般特征；两栖纲各目的主要特征；两栖纲常见科及代表种；大鲵*Andrias davidianus*、小鲵*Salamandrella*、蟾蜍*Bufo*、青蛙*Rana*等的主要特征。

● 脊柱分化特征及其与附肢骨的关系。

● 三腔心脏及不完全双循环。

14-6 巩固测验

【名词】

咽式呼吸、不完全双循环、耳柱骨、犁鼻器、完全变态、休眠

【选择】

1. 青蛙成体的呼吸器官是（　　）。

A. 肠、肺　　B. 褶鳃、肺

C. 肠、皮肤、肺　　D. 肺、皮肤、口咽腔黏膜

2. 蝌蚪的心脏组成是（　　）。

A. 一心房、一心室　　B. 两心房、两心室

C. 一心房、两心室　　D. 两心房、一心室

3. 两栖类成体的血液循环属于（　　）。

A. 单循环　　B. 完全的双循环

C. 开管式循环　　D. 不完全的双循环

4. 两栖类的大脑表皮为（　　）。

A. 原脑皮　　B. 古脑皮

C. 上皮组织　　D. 大脑皮层

5. 蛙类的耳柱骨来源于（　　）。

A. 方骨　　B. 关节骨

C. 舌颌骨　　D. 角舌骨

【填空】

1. 现有的两栖动物有（　　　　）、（　　　　）、（　　　　）3个目，约4200种。

2. 两栖类的脊柱发展成颈椎、（　　　　）、（　　　　）和尾椎，颈椎1枚，又称（　　　　），荐椎（　　　　）枚，尾椎骨愈合成一棒状的（　　　　）。

3. 两栖类肩带由（　　　　　）、（　　　　　）、（　　　　　）和（　　　　　）构成；腰带由（　　　　　）、（　　　　　）和（　　　　　）构成。

【简答】

1. 两栖纲对陆地环境的初步适应特征有哪些？其离不开水的特点有哪些？动物登陆面临的矛盾有哪些？

2. 两栖类在陆生生活的适应方面表现出哪些不完善性？

3. 简述蟾蜍血液循环系统的基本构成及心脏的基本结构。

4. 简述青蛙呼吸的特点。

Chapter 15 Class Reptilia

15-1 Characteristics of Reptiles

- Skin is dry, rough and without glands, bearing epidermal scales or scutes. Snakes and lizards shed their scales as skin cast.
- They do not respire by means of gills. Respiration always takes place through lungs. Ribs help to expand and contract the body cavity, making the lung respiration more efficient than that in amphibians.
- Skull is monocondylic, i.e. with single occipital condyle. Except in snakes, there are two pairs of pentadactyl limbs, each with five digits bearing claws—tetrapodus pentadactyl type.
- Heart consists of two auricles and a partially divided ventricle. In crocodilians, heart is four chambered (two auricles and two ventricles). Renal portal system is less developed. Red blood cells are nucleated.
- Kidneys are metanephric. Urinary bladder may be present. Crocodiles are ammonotelic. Turtles and alligators are ureotelic. Lizards and snakes are uricotelic.
- Twelve pairs of cranial nerves are present. Each ear consists of three parts: external, middle and internal. Snakes do not possess ears. The lateral line system is altogether absent.
- Tortoises feed almost entirely on vegetation. Some turtles are flesh eaters. All other reptiles are carnivorous/insectivorous. A typical cloaca is present.
- They are mostly oviparous. Reptiles lay macrolecithal eggs (= polylecithal eggs). Some forms are ovoviviparous or viviparous.
- During development, in reptiles, birds and mammals, embryo forms four membranes called embryonic membranes. These are chorion, amnion, allantois and yolk sac. Due to their occurrence, reptiles, birds and mammals are called amniotes. Fishes and amphibians do not have these membranes; hence they are called anamniotes.

● Four features make reptiles true land animals: internal fertilization; the amnion (embryonic membrane) encloses the embryo and provides it with a watery environment during development so that embryo does not need watery environment; shell around egg checks desiccation; horny scales on body of reptiles check loss of water.

Reptiles are the creeping and burrowing cold-blooded vertebrates bearing epidermal scales. They are ectothermic (cold-blooded) and are found mostly in the warmer parts of the world. They are few in colder parts. They are mostly terrestrial animals. There are about 6,000 living species of reptiles in the world.

● External Features

The body comprises a distinct head, neck, trunk, and tail; each of the short limbs bears toes tipped with horny claws and with webs between. The long mouth is margined with conical teeth, set in sockets (Fig. 15-1). Near the tip of the snout are two small valvular nostrils. The eyes are large and lateral, with upper and lower eyelids, and a transparent nictitating membrane moves backward beneath the lids. The ear opening is behind the eye under a movable flap. The vent is a longitudinal slit, behind the bases of the hind limbs.

● Body Covering

The tough leathery skin is sculptured into rectangular horny scales over most of the trunk and tail. The scales are generally in transverse and lengthwise rows, with furrows of softer skin between. The cornified exterior wears are replaced by additional cornified layers from the epidermis beneath. Adults have an exoskeleton of separate bony dermal plates (osteoderms) under the dorsal scales from neck to tail; these are rectangular or oval, often pitted, and some have a median keel. Some species also have osteoderms on the belly. There are two pairs of epidermal musk glands, one pair opening on each side of the under surface of the jaw, and the other pair within the cloaca.

● Skeleton

The massive skull includes a long snout, and the bones are usually pitted in old adults. The long lower jaw articulates at each side of the posterior margin of the skull on a fixed quadrate bone. Ventrally on the cranium is the long hard palate above which are the respiratory passages. The vertebral column comprises five types of vertebrae: 9 cervical, 10 thoracic, 5 lumbar, 2 sacral, and about 39 caudal. On the cervical vertebrae are short free cervical ribs; the thoracic vertebrae and sternum are connected by thoracic ribs, with cartilaginous ventral extensions, and between the sternum and pubic bones there are seven pairs of V-shaped abdominal ribs (gastralia) held in a longitudinal series by ligaments (Fig. 15-2).

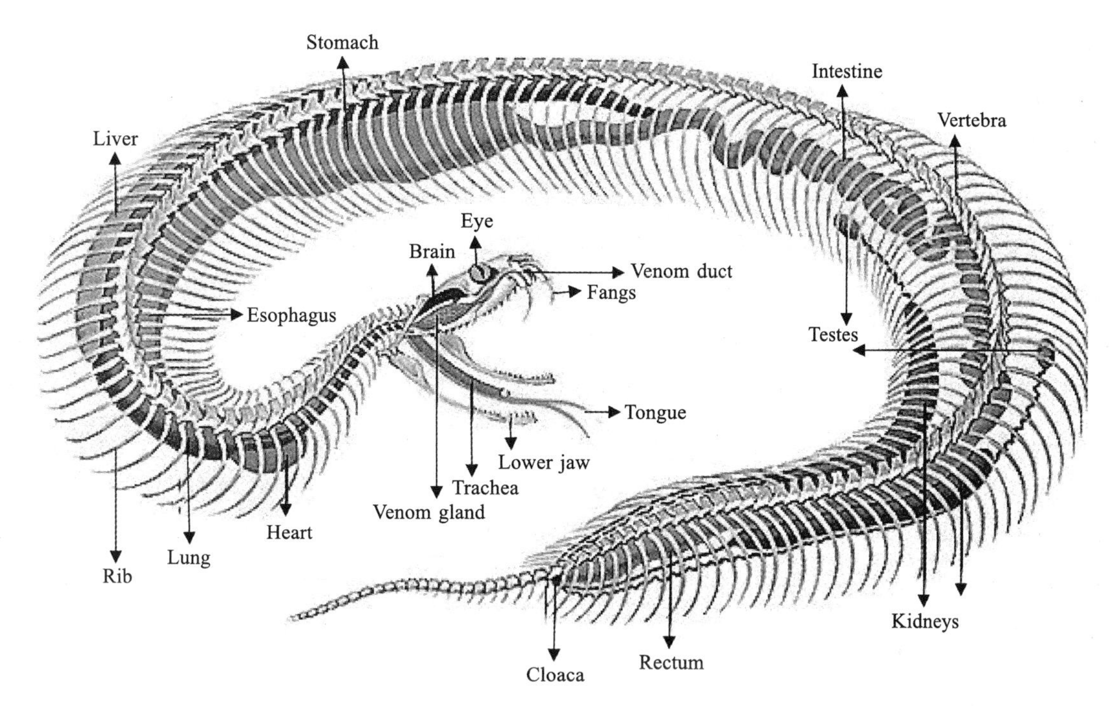

Fig. 15-1 Schematic of reptiles

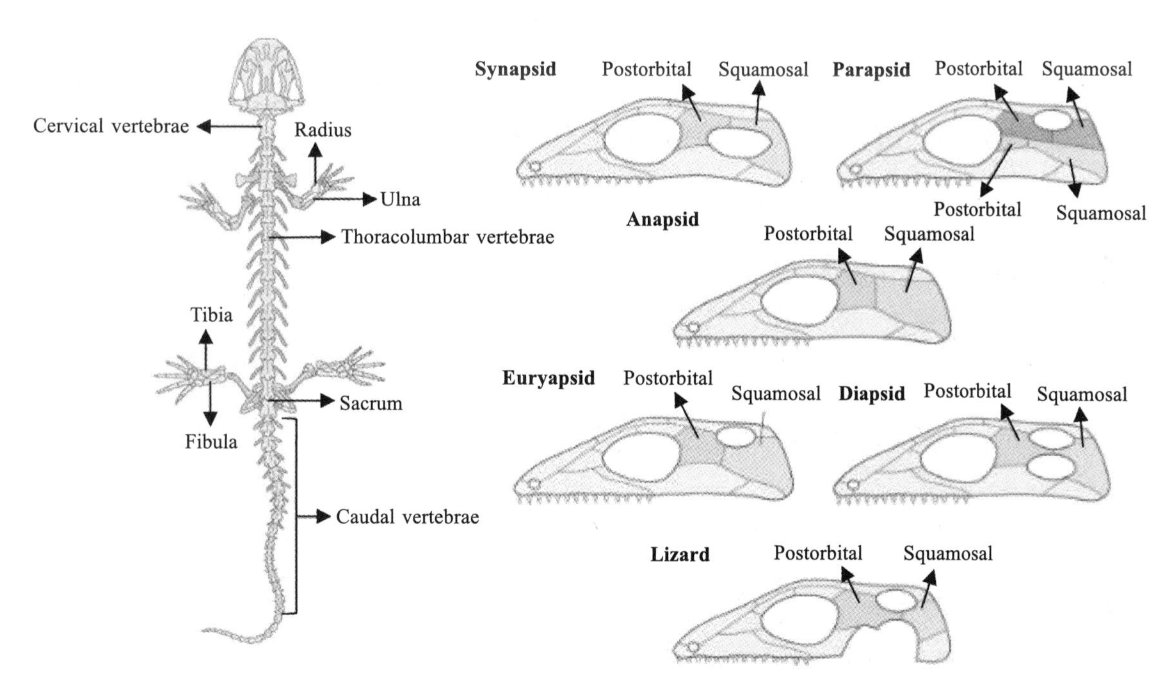

Fig. 15-2 Skeleton of reptiles

● Muscular System

As compared with a frog, the alligator has more diversity in its muscles in keeping with the greater variety of its bodily movements both on land and in water. The muscles of the head, neck, and limbs are well differentiated, though less in bulk than in a mammal. The segmental muscles of the vertebral column, ribs, and tail are conspicuous.

● **Circulatory system**

The heart lies in the anteroventral part of the thorax; it comprises a small sinus venosus, two atria, and two ventricles (Fig. 15-3). The ventricles are completely separated in crocodilians, but imperfectly in other reptiles. Blood from the veins passes in turn through the sinus venosus, right atrium, right ventricle, pulmonary artery to each lung, pulmonary veins from lungs, to the left atrium, and left ventricle. It emerges in a pair of aortic arches that pass dorsally around the esophagus; from the base of the right arch, two carotid arteries lead to the neck and head, and a subclavian artery to each forelimb. The two aortic arches join as a dorsal aorta that distributes blood to organs in the body cavity and to the hind limbs and tail. Venous blood is collected by an anterior vena cava on each side from the head, neck, and forelimb; by a single middorsal posterior vena cava from the reproductive organs and kidneys; by a hepatic portal vein from the digestive tract that breaks up into capillaries in the liver and collects as a short hepatic vein; and by an epigastric vein on each side of the abdominal cavity from the posterior limbs, tail, and body. All these veins empty into the sinus venosus.

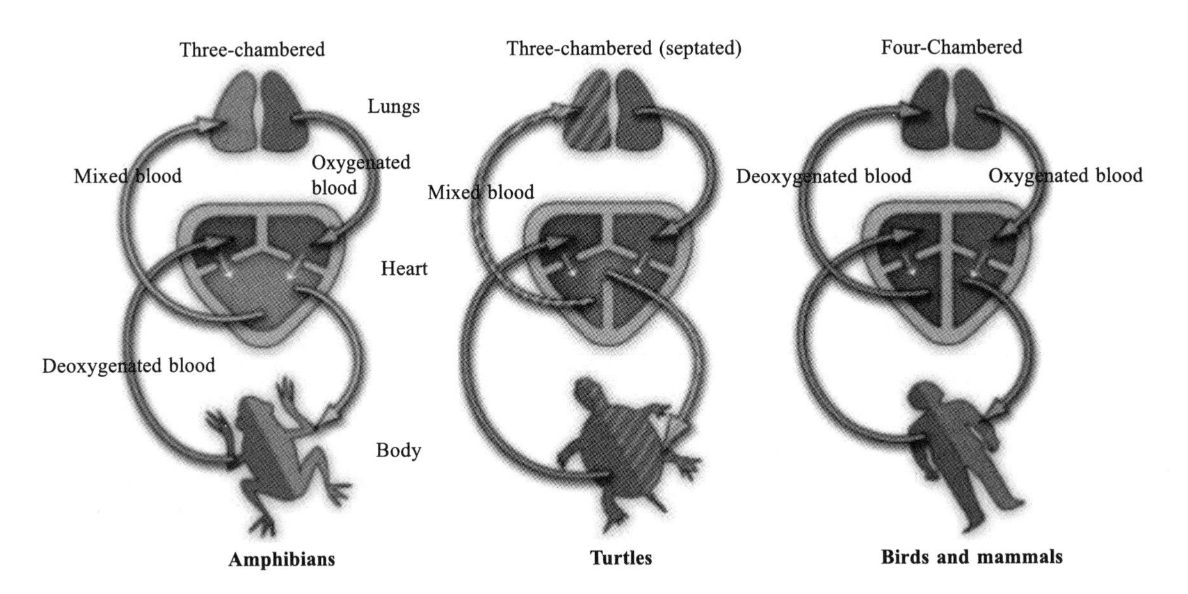

Fig. 15-3 Circulatory system of turtles in comparison with amphibians, and birds and mammals

● **Respiratory System**

Air enters the nostrils (external nares) and passes above the hard palate to the internal nares behind the palatine valve (velum) and thence through the glottis in the larynx, just behind the tongue. The larynx is supported by three cartilages and contains the paired vocal cords; it connects to the tubular trachea, which is reinforced with rings of cartilage. The trachea extends to the forepart of the thorax and divides into two short bronchi, one to each lung. The lungs contain higher interior partitions than in the frog and are spongier.

● **Excretory System**

The two flat lobular kidneys lie in the posterior dorsal part of the body cavity; a ureter from each extends back to the side of the cloaca.

● **Nervous System and Sense Organs**

The brain has two long olfactory lobes connected to the large cerebral hemispheres; behind the latter are two oval optic lobes. Next is the median pear-shaped cerebellum, which is larger than in amphibians. The medulla oblongata is spread laterally below the cerebellum, then narrows to the spinal nerve cord. Ventrally, between the bases of the cerebral hemispheres, are the optic tracts and optic nerves, followed by the infundibulum and the hypophysis. There are 12 pairs of cranial nerves and paired spinal nerves to each body somite.

There are taste buds on the tongue and olfactory cells in the nose. The eyes have lachrymal glands that keep the cornea or surface of the eyeball moist when out of water. The ears are of the type characteristic of land vertebrates. Each ear has a short external auditory canal under a flap of skin, with a tympanic membrane at the inner end, a tympanic cavity, or middle ear, housing the one ear bone, or stapes, and an inner ear containing three semicircular canals and the organ of hearing. From each tympanic cavity, two Eustachian tubes lead medially to a common opening on the roof of the pharynx behind the internal nares. In other vertebrates, the tubes are single and open separately on each side of the pharynx.

● **Structure of Amniotic Egg**

Structure of an amniotic egg is shown in Fig. 15-4.

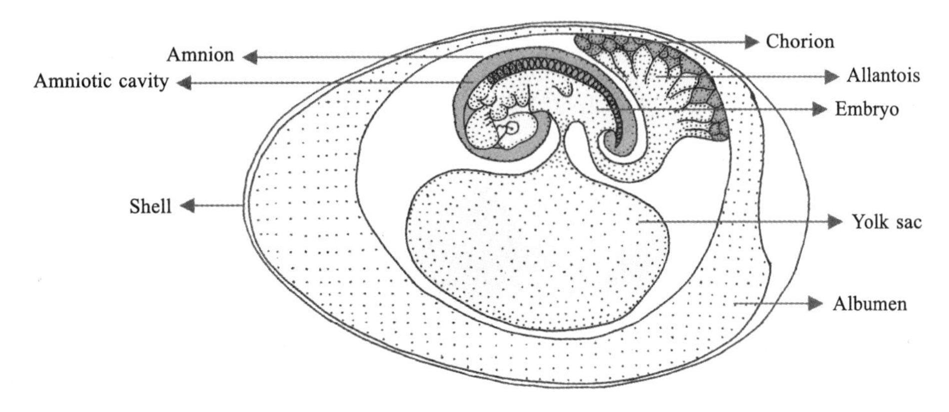

Fig. 15-4 Schematic of an amniotic egg

● **Reproductive System**

The paired gonads and ducts are much alike in the young of the two sexes. In mature males, two roundish testes lie near the ventromedial borders of the kidneys; a ductus deferens from each passes back to enter the cloaca just anterior to the ureter and joins the single median penis on the ventral floor of the cloaca. In adult females, two ovaries are similarly attached near the kidneys. Near the anterior end of each kidney is the open funnel of an oviduct, and the latter runs backward to the cloaca. Eggs form in the ovaries and pass into the funnels; in the oviducts, each is fertilized and covered with albumen, shell membranes, and a shell before being laid.

15-2 Classifications of Living Reptiles

● **Anapsida**

Skull has a solid bony roof, no temporal vacuities. Anapsida includes only single living order chelonia, e.g. *Chelone* (turtle), *Testudo* (tortoise), *Trionyx* (terrapin)—soft shelled turtle of Indian rivers (Fig. 15-5).

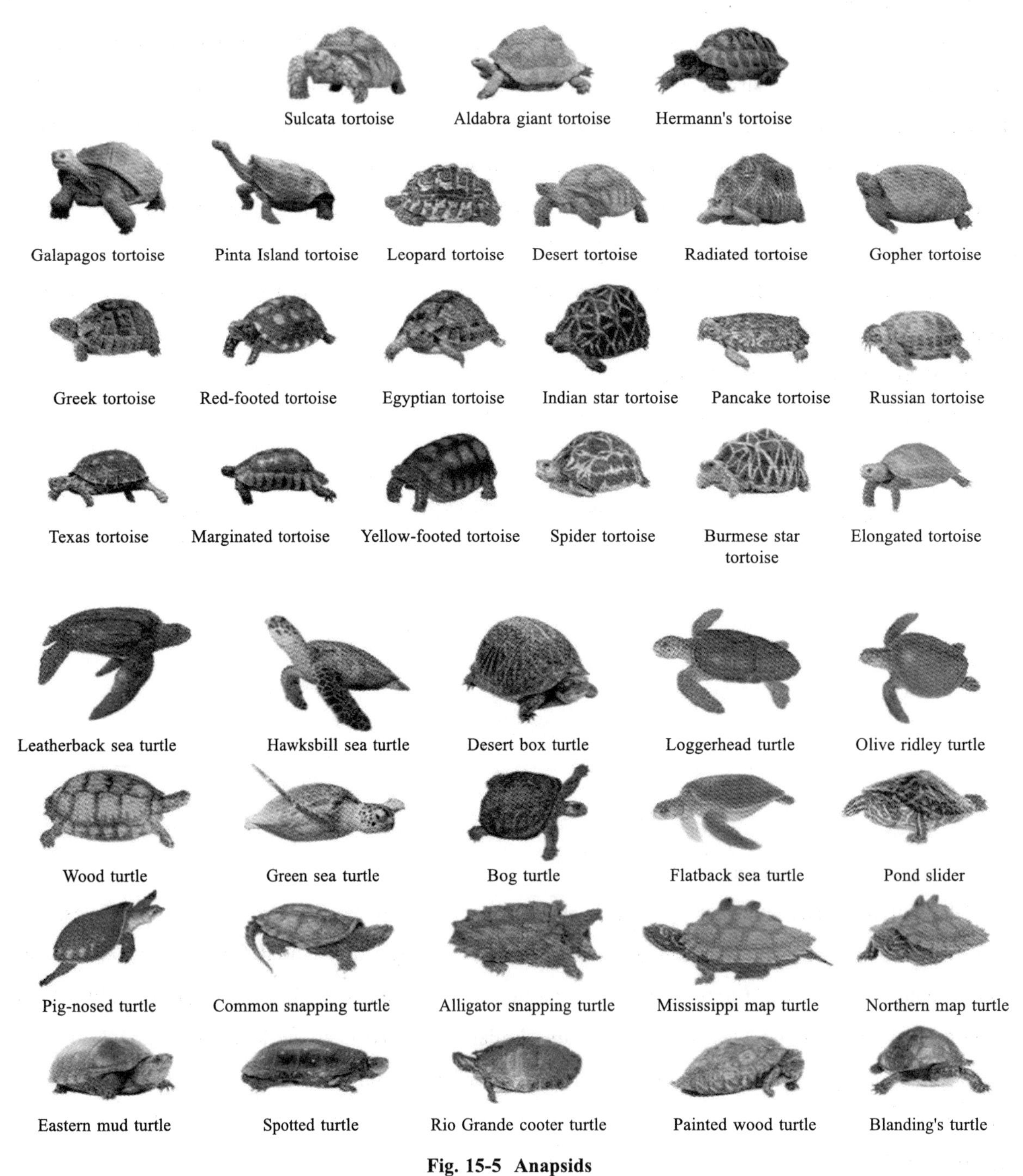

Fig. 15-5 Anapsids

● **Diapsida**

Skull has two temporal vacuities. Diapsida includes three living orders.

(1) Rhynchocephalia

Example: *Sphenodon* (tuatara)—a living fossil (Fig. 15-6).

Fig. 15-6 Tuatara

(2) Squamata

It includes two suborders:

Suborder Lacertilia (Sauria): e.g. lizards, such as chameleon (tree lizard), *Calotes* (garden lizard), *Hemidactylus* (wall lizard) (Fig. 15-7).

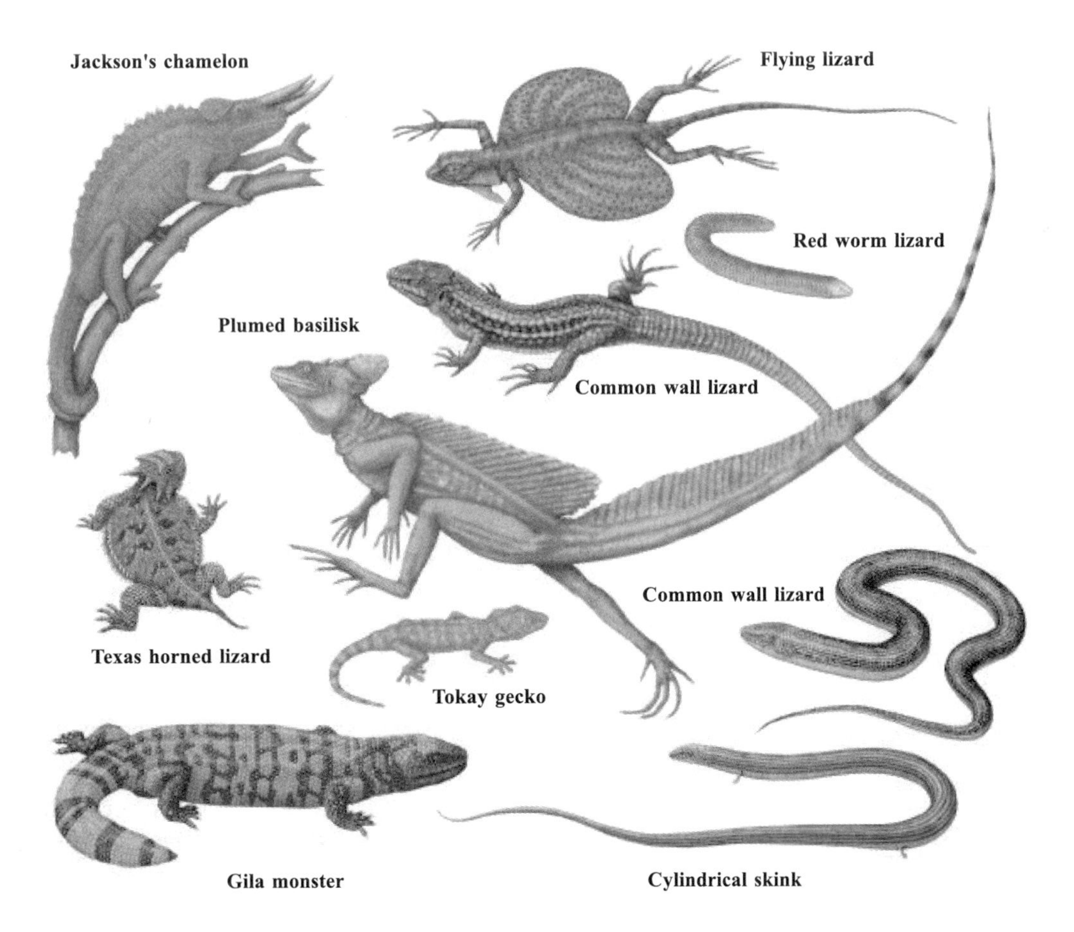

Fig. 15-7 Lizards

Suborder Ophidia: e.g. snakes, such as *Naja* (cobra), *Bungarus* (krait), *Vipera* (viper) (Fig. 15-8).

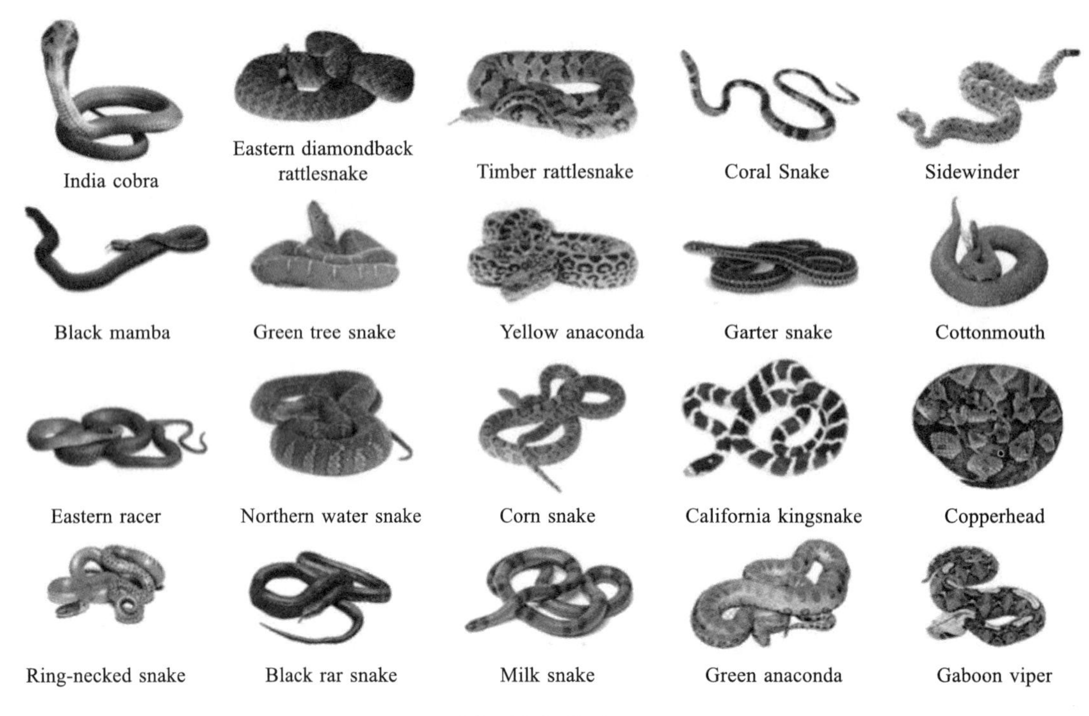

Fig. 15-8 Snakes

(3) Crocodilia

They have thecodont teeth, lungs in pleural cavities, a muscular diaphragm, analogous to that of mammals and four-chambered heart.

Examples: *Crocodylus* (crocodile), Alligator, *Gavialis* (gharial) (Fig. 15-9).

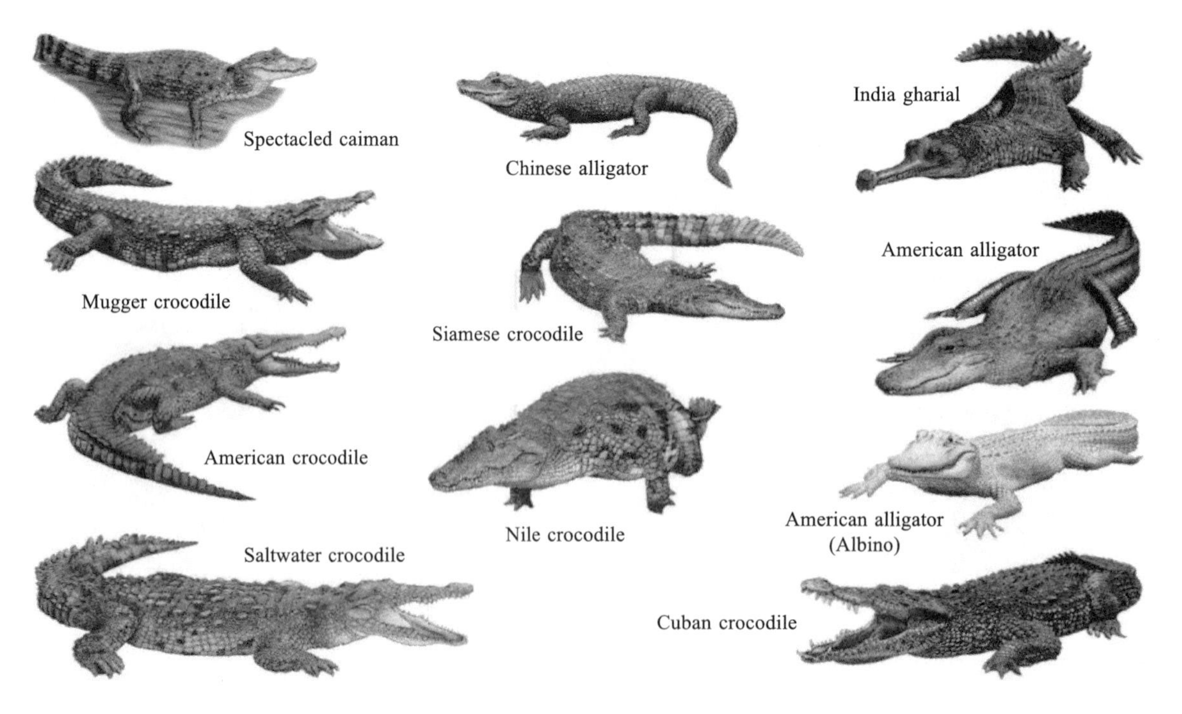

Fig. 15-9 Crocodilians

15-3 Extinct Groups of Class Reptilia

The following extinct groups of class Reptilia are important to mention here (Fig. 15-10).

● **Cotylosauria**

They were most primitive reptiles and closest to early amphibians. They were without temporal fossae in the skull, e.g. *Seymouria*.

● **Ichthyopterygia**

They were fish-like and had single fossa in the skull, e.g. *Ichthyosaurus*.

● **Archosauria**

They had diapsid skulls. Some were bipedal and gave rise to birds. A group of Archosauria also

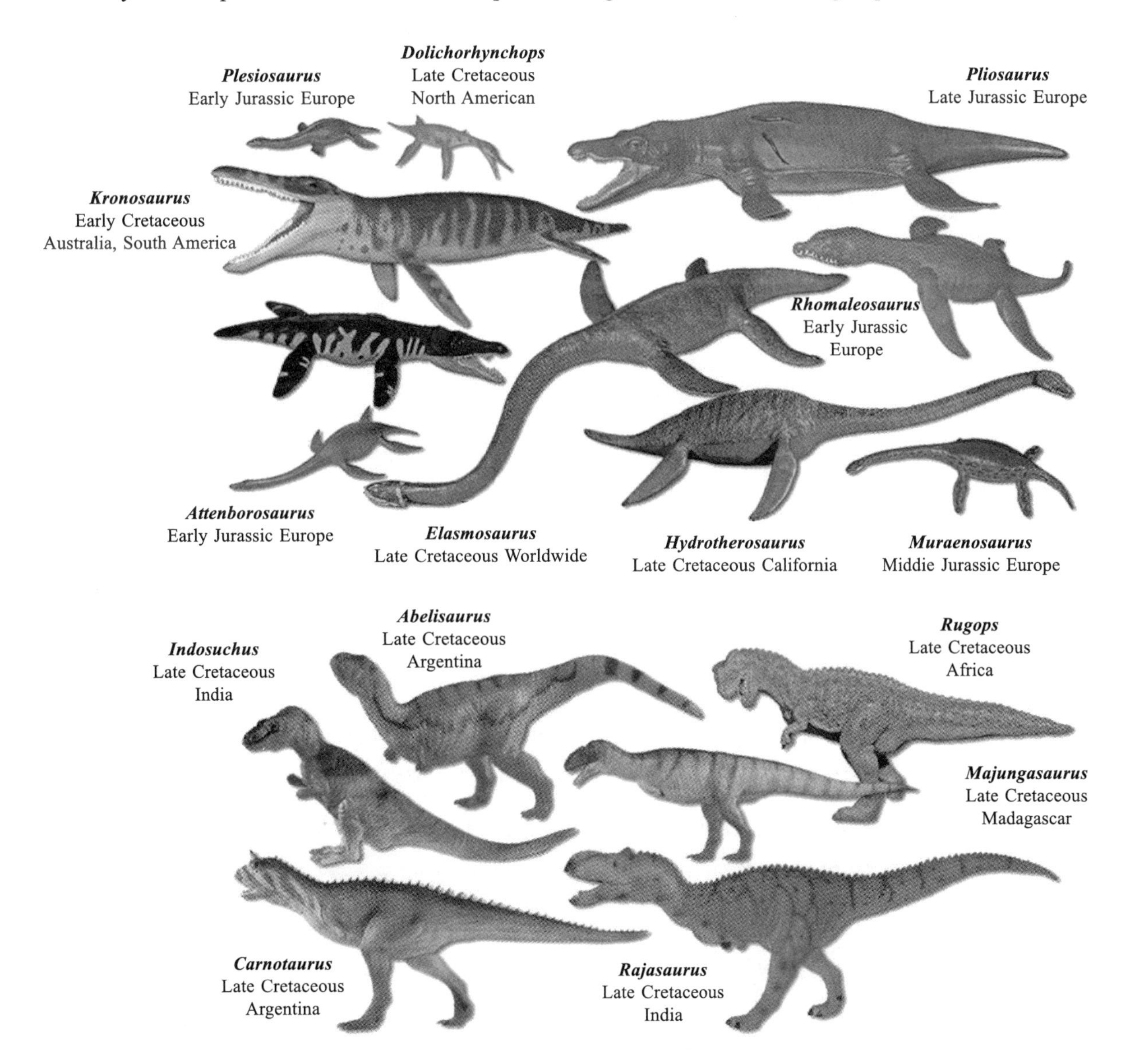

Fig. 15-10 Extinct reptiles

gave rise to dinosaurs, e.g. *Brontosaurus*.

- **Synaptosauria**

The skull had a single temporal fossa on either side. They were mammal-like reptiles that later on gave rise to mammals, e.g. *Plesiosaurus*.

15-4 学习要点

- 羊膜卵的形成、构造及在脊椎动物演化史上的意义。
- 爬行类对陆地生活的完全适应性。
- 爬行类躯体结构的基本特征；爬行纲的分类；龟鳖目各科的特点及经济意义；鳖 *Peldiscus sinensis*、乌龟 *Chinemys* 等的特征及经济意义。
- 颞孔。
- 胸腹式呼吸的特征。
- 爬行类中轴骨的结构。
- 爬行类循环系统的结构特征。

15-5 巩固测验

【名词】

亚卵胎生、不完全双循环、羊膜卵、潘氏孔、次生腭、胸腹式呼吸、颞孔、原脑皮与新脑皮

【选择】

1. 具有一个枕髁的动物是（　　）。

A. 两栖类和爬行类　　B. 鸟类和爬行类

C. 鸟类和哺乳类　　D. 两栖类和哺乳类

2. 腰带由（　　）构成。

A. 髂骨、坐骨、耻骨　　B. 肩胛骨、乌喙骨、锁骨

C. 髂骨、坐骨、尾骨　　D. 坐骨、耻骨、荐椎、尾骨

3. 爬行类从体形上显著区别于两栖类的特征是（　　）。

A. 具有五趾型附肢　　B. 有尾

C. 皮肤富有腺体　　D. 皮肤干燥，有角质鳞片或盾片，指/趾端有爪

4. 除了肺呼吸外，还有副膀胱辅助呼吸的爬行动物是（　　）。

A. 石龙子　　B. 蛇类

C. 壁虎　　D. 龟鳖

5. 犁鼻器是（　　）器官。

A. 嗅觉　　B. 听觉

C. 红外线感受器　　D. 压力感受器

6. 爬行类皮肤的最大特点是（　　）。

A. 干燥，被骨质鳞　　B. 缺乏皮脂腺，被骨板

C. 干燥，被角质鳞　　D. 缺乏皮脂腺，被盾鳞

7. 爬行类的大脑皮层为（　　）。

A. 原脑皮　　B. 后脑皮

C. 大脑皮层　　D. 新脑皮

【简答】

1. 为什么说爬行类是真正的陆生脊椎动物？

2. 简述羊膜卵的结构及其在脊椎动物演化史上的重要意义。

3. 颞孔指什么？它有哪些类型？

Chapter 16 Class Aves

16–1 Characteristics of Aves

- Birds are warm-blooded animals. Their forelimbs are modified into wings. The hindlimbs are adapted for walking, hopping, grasping, perching, wading and swimming. There is no skin gland except the oil gland. The upper and lower jaws are modified into a beak. They have no teeth. There are epidermal scales on their legs.
- The alimentary canal has a crop and a gizzard. The crop helps in softening food, and the gizzard helps in crushing food. Some seed-eating birds do not have a gall bladder.
- They have spongy and elastic lungs for respiration. The voice is produced by a special organ—the syrinx.
- The heart is four-chambered. The red blood cells are nucleated, oval, and biconvex.
- They have 12 pairs of cranial nerves. They have a sharp sight.
- The endoskeleton is bony with long hollow bones filled with air cavities.
- They have a single ovary and oviduct on the left side.
- All the birds are oviparous and exhibit sexual dimorphism. The eggs have four embryonic membranes—chorion, amnion, allantois, and yolk sac.
- They have a spindle-shaped body to minimize resistance to the wind.
- The feathers allow the air to pass and reduce friction to a minimum.
- They have well-developed flight muscles that help during flight.

The Aves belong to the phylum Chordata of the animal kingdom. It has about 9,000 species. Aves are adapted to fly. All the birds come in the class Aves. Members of this class exhibit one of the most beautiful and visually stunning features such as bright and contrasting colors, unique patterns and a wide variety of striking poses. They show courtship, parental care, nest building, and territorial behavior.

- **External Features**

The chicken or fowl has a distinct head, a long flexible neck, and a stout spindle-shaped body, or trunk. The two forelimbs, or wings, are attached high on the back and have long flight feathers (remiges); the wings are deftly folded in Z-shape at rest and extended in flight. On each hindlimb the two upper segments are muscular, whereas the slender lower leg, or shank, contains tendons but little muscle and is sheathed with cornified scales, as are the four toes, which end in horny claws (Fig. 16-1). The short tail bears a fanlike group of long tail feathers (rectrices) (Fig. 16-2).

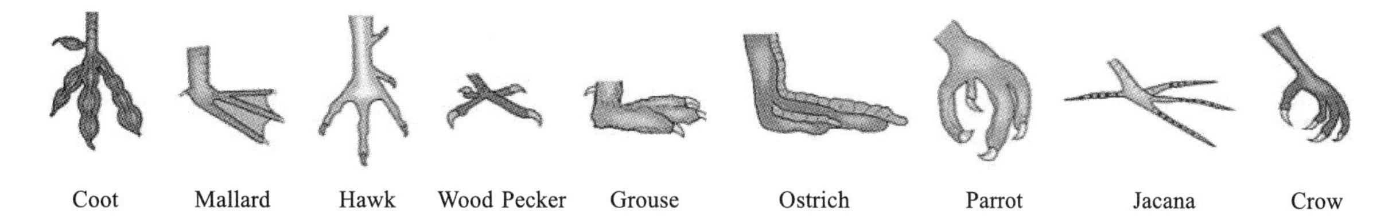

Fig. 16-1 Toes and claws

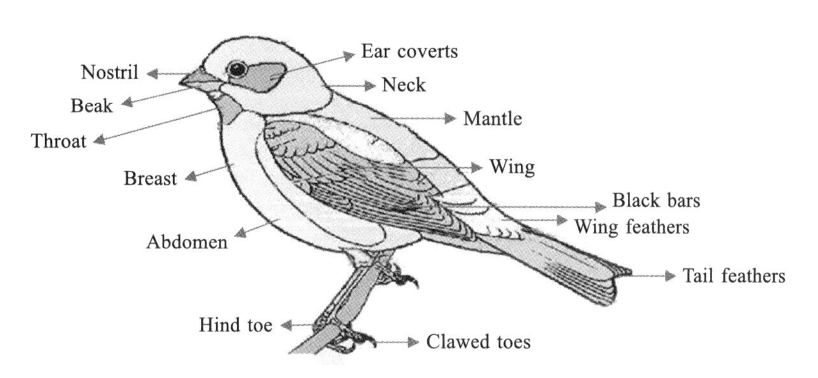

Fig. 16-2 Schematic of birds

The mouth is extended as a pointed bill, or beak, with horny covering (Fig. 16-3). On the upper mandible are two slitlike nostrils. The eyes are large and lateral, each with an upper and lower eyelid; beneath these is the membranous nictitating membrane, which can be drawn independently across the eyeball from the anterior corner. Below and behind each eye is an ear opening, hidden under special feathers. The fleshy median comb and lateral wattles on the head and the cornified spurs on the legs are peculiar to the chicken, pheasant, and a few other birds. Below the base of the tail is the vent.

Fig. 16-3 Beaks

● **Body Covering**

The soft, flexible skin is loosely attached to the muscles beneath. It lacks glands save for the oil gland above the base of the tail, which secretes an oily substance for "dressing" the feathers and to keep the bill from becoming brittle. The feathers grow from follicles in the skin.

● **Feathers**

These distinctive epidermal structures provide a lightweight, flexible, but resistant body covering, with innumerable dead air spaces useful as insulation; they protect the skin from wear, and the thin, flat, and overlapping wing and tail feathers form surfaces to support the bird in flight (Fig. 16-4).

Fig. 16-4 Feathers of birds

(1) Contour Feathers

These provide the external covering and establish the contour of the bird's body, including the enlarged flight feathers of the wings and tail; several thousands are present on a chicken. Each consists of a flattish vane, supported by the central shaft, which is an extension of the hollow quill attaching to the follicle. Each half of the vane is composed of many narrow, parallel, and closely spaced barbs joining the sides of the shaft. On the proximal and distal side of each barb are numerous smaller parallel barbules, and these are provided with minute barbicels, or hooklets, serving to hold opposing rows of barbules loosely together. Many body feathers have a secondary shaft, or aftershaft, and vane, attached to the junction of the principal shaft and quill. Smooth muscles and elastic fibers in the skin enable a bird to ruffle or raise its feathers away from the body, to facilitate their rearrangement when bathing and preening, and to increase the insulation value of the feather covering during cold weather.

(2) Down Feathers

Young chicks and many other birds at hatching are covered with soft downy plumage, providing excellent insulation. A down feather has a short quill, a reduced shaft, and long flexible barbs with short barbules. Down is also present beneath the contour feathers on ducks, many other water birds, and some land birds.

(3) Filoplumes

Minute hairlike feathers, of unknown function, are sparsely distributed over the body, as seen on a plucked fowl. These grow in clusters near the follicles of some contour feathers; each has a long threadlike shaft, with a few weak barbs and barbules at the tip.

(4) Bristles

Some birds have hairlike growths that are modified feathers, each with a short quill and slender shaft, with a few vestigial barbs at the base. These are seen about the mouths of flycatchers and whippoorwills and serve a tactile function.

● **Skeleton**

The bird skeleton is delicate as compared with that of most mammals; many of the bones contain air cavities to lessen the weight. The skeleton is modified in relation to flight, bipedal locomotion, and the laying of large eggs with hard shells. The bones of the cranium are separate in young birds but fused in adults except for a nasofrontal hinge that permits slight movement of the upper jaw in many species. The brain case is rounded, the orbits sheltering the eyes are large, and the jaws (premaxillae, maxillae, and mandibles) project forward as the bony beak. The lower jaws hinge on the movable quadrate connecting to the squamosal. The skull articulates on a single occipital condyle with the first neck vertebra (Fig. 16-5).

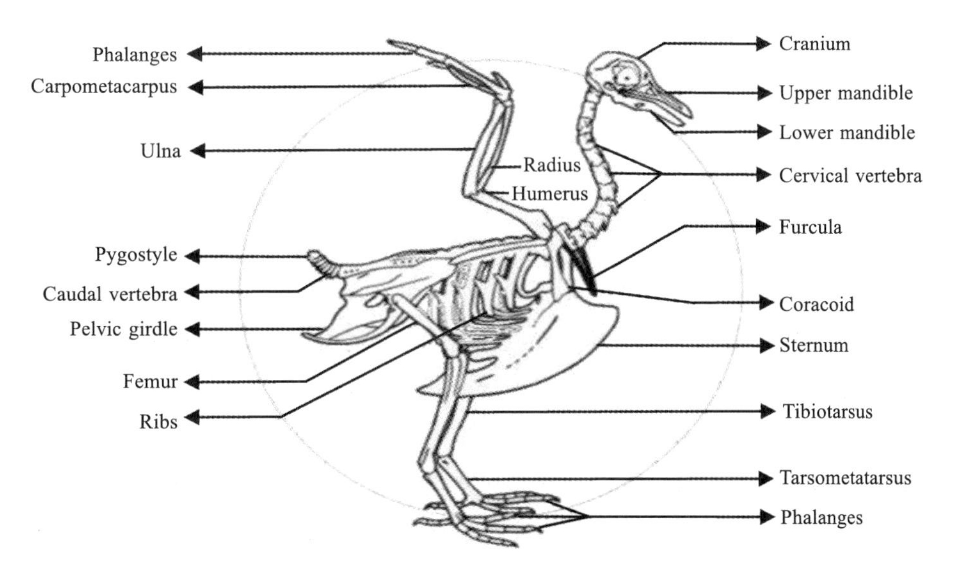

Fig. 16-5 Skeleton of birds

The neck contains about 16 cervical vertebrae, each with saddle-shaped bearing surfaces that permit free movements in feeding, preening, and other activities. The trunk vertebrae are closely fit together; those of the thorax have rib articulations laterally, and the remainders are fused into a solid synsacrum to which the pelvis attaches. No lumbar region is evident. The four free caudal vertebrae and the compressed terminal pygostyle (five or six fused vertebrae) serve in movements of the tail feathers. The bony thorax protects the internal organs and provides a rigid support for the flight mechanism, yet is capable of slight expansion and contraction for respiration.

The pectoral girdle on each side consists of the swordlike scapula (shoulder blade) lying parallel to the vertebrae and over the ribs, the coracoid as a stout stay between the scapula and sternum, and the clavicle hanging vertically from the scapula; the two clavicles are fused at their ventral cnds to form the V-shaped furcula, or "wishbone", attaching to the sternum. The three bones meet dorsally on either side to form a circular canal as a pulley for the tendon of the supracoracoideus muscle that lifts the wing (Fig. 16-6). Each forelimb attaches high on the dorsal surface, the humerus pivoting in the glenoid fossa on the coracoid. The pelvic girdle is a broad, thin saddle firmly united to the synsacrum but widely open ventrally, permitting easy passage of large eggs in the female.

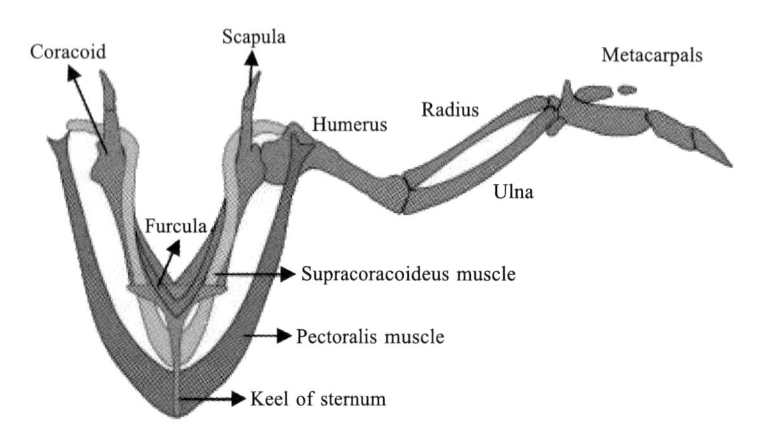

Fig. 16-6 Pectoral girdle of birds

Each leg consists of the femur, or thigh-bone; the long and triangular-headed tibiotarsus, which is paralleled by the slender and often incomplete fibula; the fused tarsometatarsus, or shank; a knee bone, or patella, held in ligaments before the femur-tibiotarsal joint; and the four toes, three in front and one behind, each consisting of two to five bones. Of the tarsals, or ankle bones, seen in other vertebrates, the proximal row in birds is fused to the tibia and the distal row to the metatarsus; the latter consists of three fused metatarsals, distinguishable only at the ends of the bone.

● **Muscular System**

In the bodies of lowest vertebrates, segmental muscles predominate over nonsegmental ones, but the reverse is true among birds and mammals, where the limb muscles are enlarged for rapid activity. Movement of the wings in flight is due chiefly to the large pectoral muscles of the breast, the "white meat" of chickens and turkeys, a major fraction of the entire musculature. At either side, the pectoralis originates on the outer part of the sternal keel and inserts on the ventrolateral head of the humerus; its contraction moves the wing downward and lifts the bird's body in flight. In other land vertebrates the forelimb is raised by muscles on the dorsal surface, but in birds such movement is also due to a ventral muscle, the supracoracoideus. This originates on the keel (inside the pectoralis) and tapers to a strong tendon passing upward to insert on the dorsoposterior surface of the humerus. Both these muscles are symmetrically paired and in turn exert an equal and opposite pull on the thin keel of the sternum. The muscles of the femur and tibiotarsus are the principal ones used for running and perching. The shanks and feet contain little muscle, an adaptation to prevent loss of heat from these unfeathered parts. The toes are moved by tendons connected to muscles in the upper segments of the legs. The tendons move through spaces lubricated by fluid, and their action on the toes is directed through loops of tendon.

● **Digestive System**

The tongue is small and pointed, with a horny covering. The mouth cavity is roofed with long palatal folds; a short pharynx follows; then the tubular and muscular esophagus extends to the base of the neck, where it dilates into a large soft-walled crop, in which food is stored and moistened. The stomach comprises a soft anterior proventriculus with thick walls secreting the gastric juices, and the disk-shaped ventriculus, or gizzard, with walls of thick, dense musculature, lined internally by hardened epithelial secretion. Here the food is ground up by action of the muscular walls, aided by bits of gravel or other hard particles swallowed for the purpose—these are, functionally, the "hen's teeth". The intestine is slender, with several coils, and leads to the larger rectum; at the junction are two slender caeca, or blind pouches.

Beyond is the dilated cloaca, the common exit for undigested food wastes and materials from the excretory and reproductive organs, ending with the vent. Dorsally, in the young, the cloacal wall bears a small outgrowth, the bursa of Fabricius, of unknown function (but useful for age determination). The large reddish liver is bilobed, with a gallbladder and two bile ducts. The pancreas usually has three ducts; all these ducts discharge into the anterior loop of the intestine (Fig. 16-7).

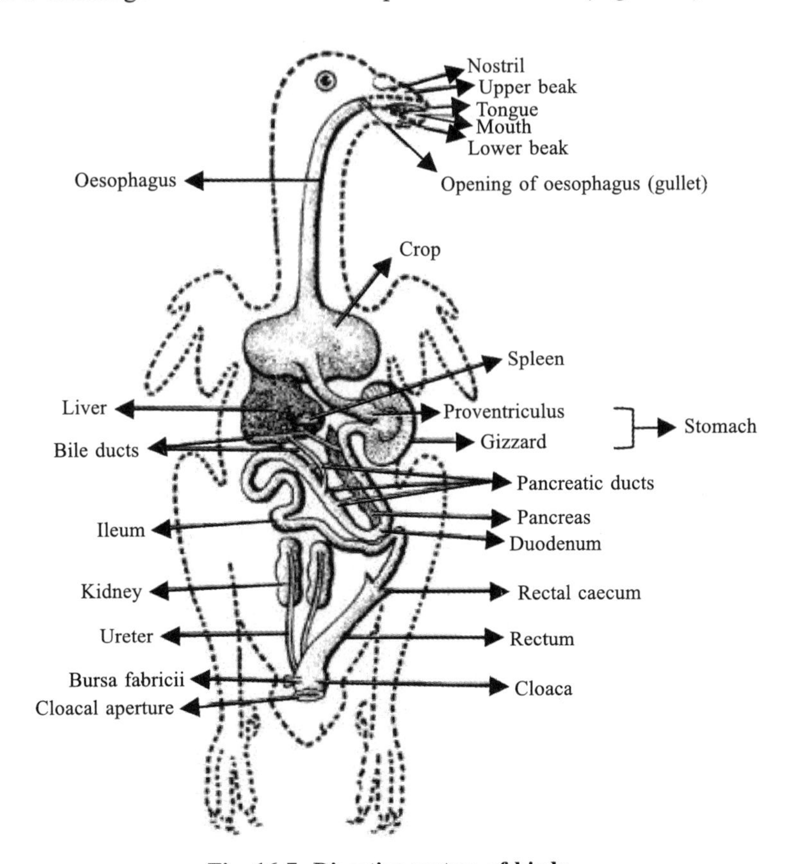

Fig. 16-7 Digestive system of birds

● **Circulatory System**

The bird heart has two thin-walled atria and two distinct thick-walled ventricles, separating completely the venous and arterial bloodstreams. This is a major factor in maintaining a regulated body temperature. The sinus venosus is incorporated in the right atrium. Blood from the two precaval veins and one postcaval enters the right atrium, passes to the right ventricle, and thence by the pulmonary artery to capillaries in the lungs for oxygenation. It all returns in the pulmonary vein to the left atrium, then the left ventricle, and into the single right aortic arch.

● **Respiratory System**

The nostrils connect to internal nares above the mouth cavity. The slitlike glottis in the floor of the pharynx opens into the long flexible trachea which is reinforced by hooplike cartilages, partly calcified. The trachea continues to the syrinx (voice box), around which are the vocal muscles. From the syrinx a bronchus leads to each lung. In a bird's lung are many small interconnecting chambers opening into larger ones, the parabronchi, which enable air to pass all the way through the lung. The parabronchi communicate with the bronchi and with the air sacs, which extend between organs in the body cavity, to spaces around neck vertebrae, and into the larger bones. Air is inspired by movements of muscles between the ribs, the sternum moving downward and the ribs bowing laterally, and by abdominal muscles; contraction of the thorax reverses the flow. On inspiration, a slug of air enters the bronchus and mostly bypasses the lung, entering the posterior air sacs. On expiration, it flows into the lung. Upon the next inspiration, as a second slug of air fills the posterior sacs, the first passes from the lung into the anterior sacs. Then with exhalation, that in the anterior sacs is expelled into the bronchus and to the outside, as the second slug of air from the lungs enters the anterior sacs (Fig. 16-8). In this way a continuous one-way flow of air is maintained through the lung. Its direction is opposite to that of blood flow, creating a countercurrent system. This explains the bird's efficient extraction of oxygen and ability to fly at high altitudes. Air sacs have a further function. They are the principal means of dissipating the heat from muscular contraction and other metabolic activities. The songs and calls of birds are produced by air forced across membranes in the walls of the syrinx, which vibrate and can be varied in tension to give notes of different pitch.

● **Bird Migration**

The word "migration" has come from the Latin word "migrara" which means going from one place to another. Many birds have the inherent quality to move from one place to another to obtain the advantages of the favorable condition. In birds, migration means two-way journeys—onward journey from the "home" to the "new" places and back journey from the "new" places to the "home". This movement occurs during the particular period of the year and the birds usually follow the same route. There is a sort of "internal biological clock" which regulates the phenomenon. All birds do not migrate, but all species are subject to periodical movements of varying extent. The birds which live in northern part of the hemisphere have greatest migratory power.

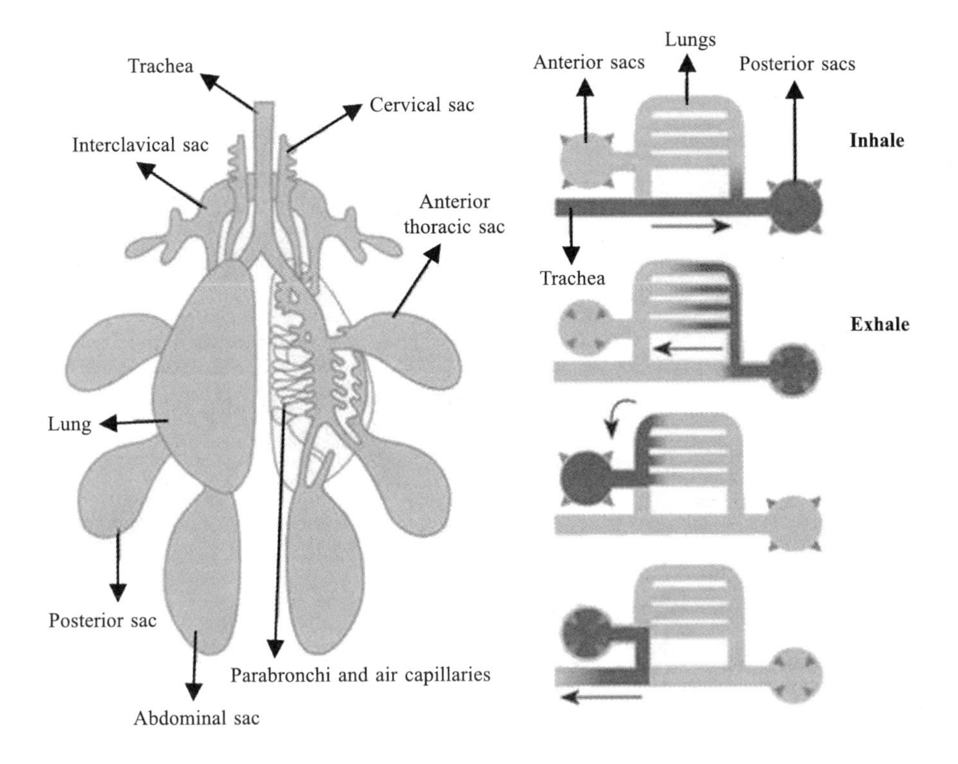

Fig. 16-8 Respiratory system of birds

16-2 Classifications of Aves

The class Aves is divided into two categories:

- **Archaeornithes**

This class of birds are extinct. They had a toothed beak with a long lizard-like tail.

Example: *Archaeopteryx*.

- **Neornithes**

These include extant as well as extinct birds. They have no teeth and a short tail.

Example: penguin, grey heron, kingfisher, duck, etc.

Orders of birds are shown in Fig. 16-9.

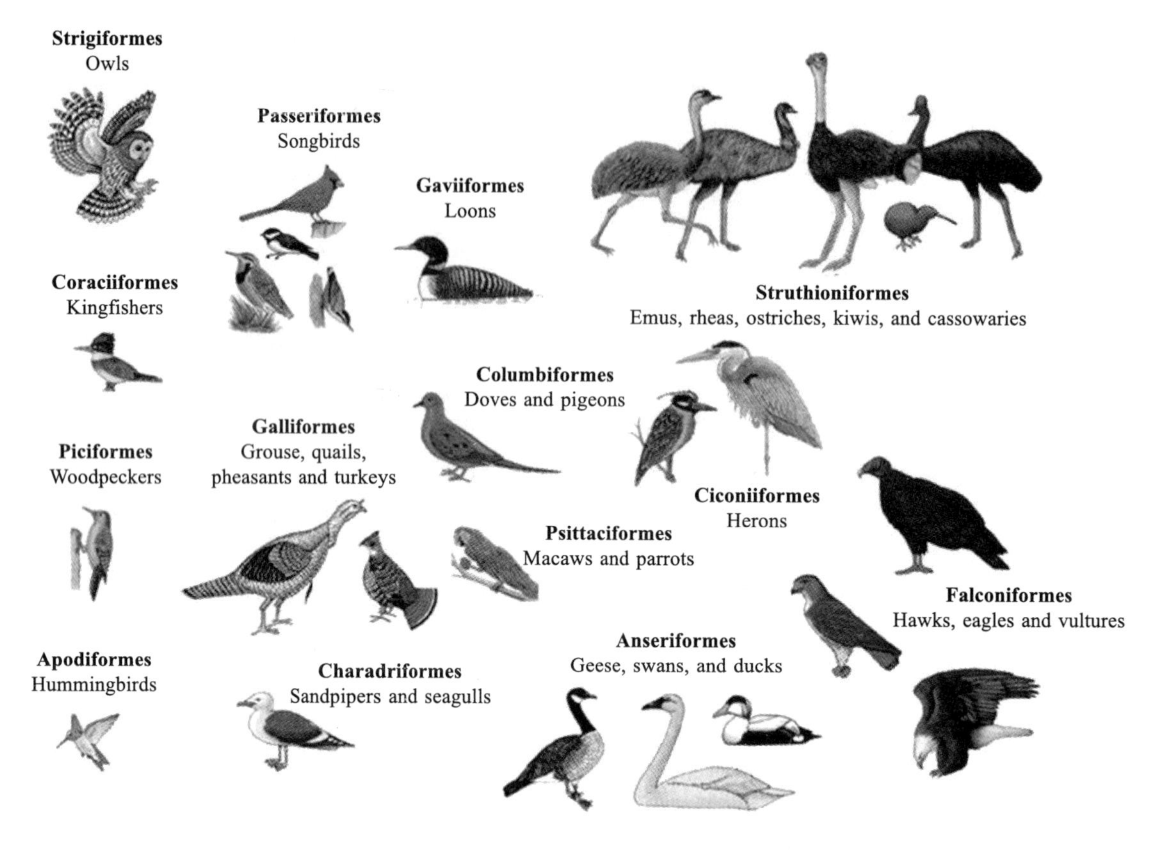

Fig. 16-9 Common orders of birds

16-3 学习要点

- 鸟类的恒温机制及在脊椎动物演化史上的意义。
- 鸟类对飞翔生活的适应性。
- 鸟类躯体结构的基本特征；鸟类的分类，突胸总目各大生态类群的结构特点及适应性。
- 呼吸器官的结构；双重呼吸。
- 鸟类视力双重调节的机制。
- 鸟类循环系统的特点。

16-4 巩固测验

【名词】

双重呼吸、双重调节、早成雏、晚成雏、迁徙、候鸟、羊膜卵、气囊

【选择】

1. 从（　　）开始出现恒温。

A. 两栖动物　　B. 爬行动物　　C. 鸟类　　D. 哺乳类

2. （　　）的尾脂腺最为发达。

A. 鸵鸟　　B. 鸣禽

C. 涉禽　　D. 水禽

3. （　　）的保温功能最好。

A. 正羽　　B. 绒羽

C. 纤羽　　D. 正羽和绒羽

4. 鸟类具有（　　）动脉弓。

A. 左体　　B. 右体

C. 左体和右体　　D. 上述选项都不正确

【简答】

1. 恒温在脊椎动物演化史上有什么意义?

2. 鸟类适应飞翔生活的主要特征有哪些?

Chapter 17 Class Mammalia

17-1 Characteristics of Mammals

- Body usually covered with hair (scant on some), which is molted periodically; skin with many glands (sebaceous, sweat, scent, and mammary).
- Skull with two occipital condyles; neck vertebrae usually seven; tail (when present) usually long and mobile.
- Nasal region usually slender; mouth with teeth (rarely absent) in sockets or alveoli, on both jaws, and differentiated in relation to food habits; tongue usually mobile; eyes with movable lids; ears usually with external fleshy pinnae.
- Four limbs (cetaceans and sirenians lack hindlimbs); each foot with five (or fewer) toes and variously adapted for walking, running, climbing, burrowing, swimming, or flying; toes with horny claws, nails, or hoofs and often fleshy pads.
- Heart completely four-chambered (two atria, two distinct ventricles); only the left aortic arch persists; red blood cells nonnucleated, usually circular.
- Respiration only by lungs; larynx with vocal cords (except in giraffes); a complete muscular diaphragm separating lungs and heart from abdominal cavity.
- A urinary bladder; excretions (urine) fluid.
- Twelve pairs of cranial nerves; brain highly developed, both cerebrum and cerebellum large.
- Homoiothermal.
- Male with copulatory organ (penis); testes commonly in a scrotum external to abdomen; fertilization internal; eggs usually minute, without shells, and retained in uterus (modified oviduct) of female for development; embryonic membranes (amnion, chorion, and allantois) present; usually with a placenta affixing embryo to uterus for nutrition and respiration; young nourished after birth by milk secreted from mammary glands of female.

Animals belonging to class Mammalia are referred to as mammals. Mammals are one of the most evolved species in the animal kingdom and are categorized under Vertebrata. They exhibit advanced characteristics which set them apart from all other animals. They are characterized by the presence of mammary glands through which they feed their younger ones. They are distributed worldwide and have adapted well to their surroundings—from oceans, deserts and polar regions to rainforests and rivers etc.

● **External Features**

The entire body is usually densely covered with hair, or fur, and consists of a rounded head, short neck, narrow trunk, and long flexible tail. Each forelimb has five toes provided with fleshy pads and curved retractile claws. The hindlimbs are stouter, providing the principal power in locomotion.

● **Body Covering**

The skin is soft and thin except for the thickened cornified pads of the feet. It is covered densely with closely spaced hairs, and all the hairs collectively are termed the coat, or pelage (Fig. 17-1).

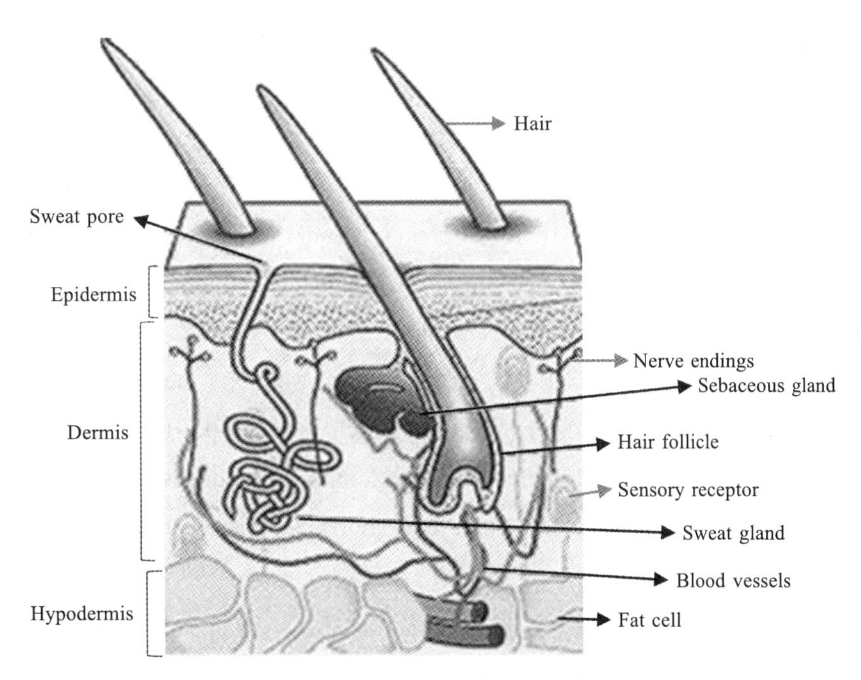

Fig. 17-1 Body covering of mammals

● **Skeleton**

The skeleton is largely of bone, with cartilage over joint surfaces, on parts of the ribs, and in a few other places. Besides the cartilage and membrane bones, certain tendons contain ossifications known as sesamoid bones; the most conspicuous of these is the kneecap, or patella, but others occur on the feet. The rounded skull is a hard case with all the bones closely united by irregular sutures that maybe obliterate in later life. The facial region contains, dorsally, the nostrils and the large orbits that shelter the eyes, and ventrally there is a flat palate margined by the teeth of the upper jaw. Outside of each orbit is a conspicuous horizontal bar, the zygomatic arch. On the posterior surface is the large foramen magnum through which the nerve cord connects to the brain, and at either side of this is a rounded occipital condyle by which the skull articulates to the first vertebra, or atlas. The lower jaw, which also bears teeth, consists of a single bone on each side that articulates to the squamosal bone of the cranium (Fig. 17-2).

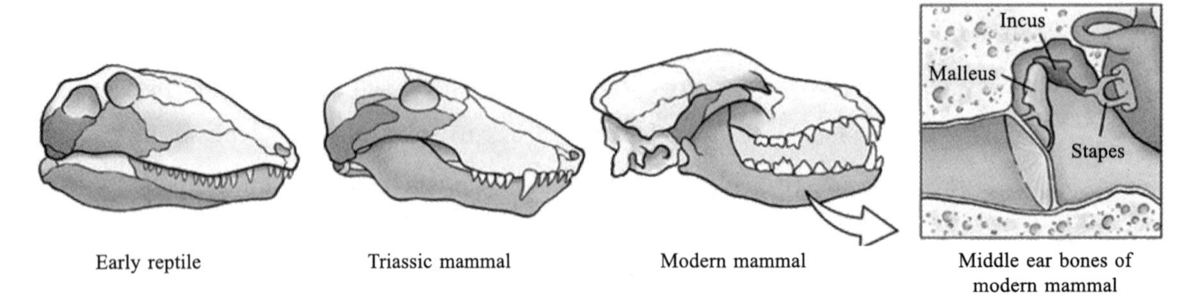

Fig. 17-2 Skull of mammals in comparison with reptiles

The vertebral, or spinal, column forms a flexible support for the body and shelters the nerve cord; adjacent vertebrae are separated by intervertebral disks of dense fibrocartilage. The column comprises five regions: the short neck of 7 cervical vertebrae, the 13 thoracic vertebrae on which the movable ribs articulate, the 7 lumbar vertebrae of the lower back, the 3 sacral vertebrae which are fused for attachment of the pelvic girdle, and the 16 to 20 slender caudal vertebrae in the tapered tail. The 13 pairs of ribs and the slender midventral sternum form a flexible "thoracic basket" that protects vital organs within and also performs respiratory movements.

The pectoral girdle attaches by muscles to the thorax and supports the forelimbs. On each side it comprises a flat triangular shoulder blade, or scapula, which receives the head of the humerus, and a delicate collarbone, or clavicle, in nearby muscles. The forelimb comprises a humerus, distinct radius and ulna, seven carpal bones, five metacarpals (innermost short), and the phalanges of the toes (Fig. 17-3).

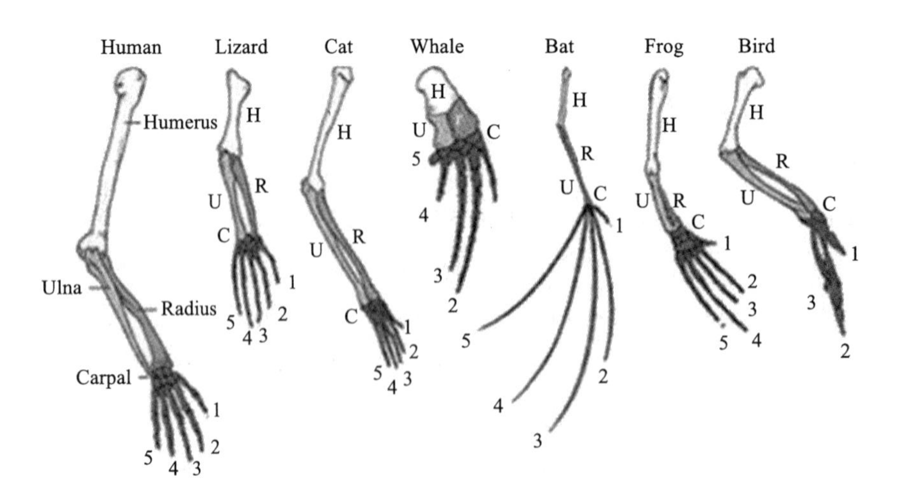

Fig. 17-3 Pectoral girdles

The pelvic girdle attaches rigidly to the sacrum; each half (the innominate bone, or hipbone) consists of an anterior dorsal ilium, a posterior ischium with a tuberosity on which the cat sits, and the ventral pubis. At the junction of the three bones is a cuplike socket, or acetabulum, in which the head of the femur articulates. The two pubes and ischia join in symphyses ventrally below the vertebrae. The bones of each hindlimb are the femur, separate tibia and fibula, seven tarsals of the ankle, four long metatarsals and a vestige of the innermost, or first, and the phalanges.

● **Teeth**

Mammalian teeth are fixed in sockets and of definite number. In various mammals the teeth are specialized in form and function according to the kind of food used (Fig. 17-4). On each tooth the exposed part, or crown, is covered by hard white enamel over a bonelike dentine that contains a pulp cavity. The base, or root, of the tooth, below the gums, is fixed by bony cement in a socket, or alveolus, the jaw. Incisors, canines, and premolars comprise the "milk teeth" of a kitten; these later are all replaced, and there are added molars having no milk predecessors. The teeth are alike on the two sides but differ in the upper and lower jaws. All the teeth collectively form the dentition, and their number is expressed as a dental formula that indicates those of the upper and lower jaws on one side.

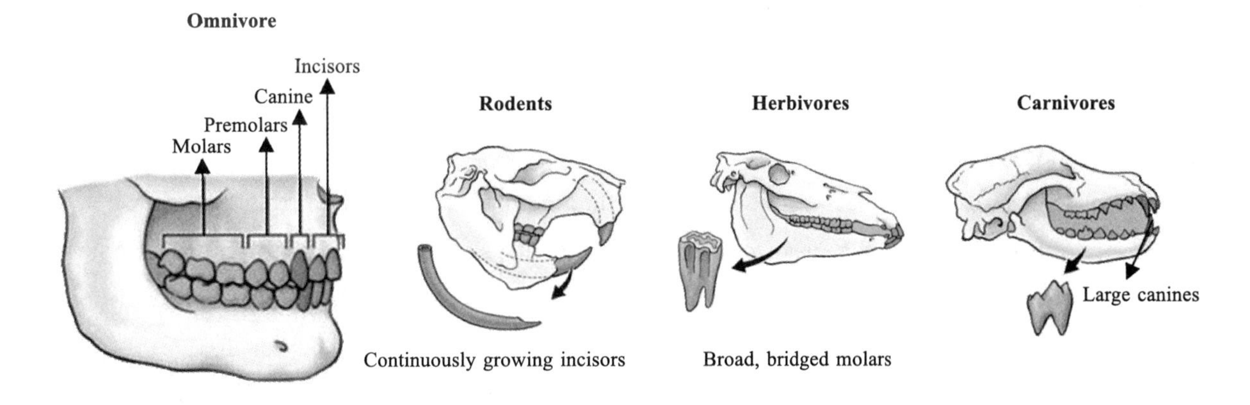

Fig. 17-4 Specialized teeth in mammals

Here are some types of mammalian skulls and teeth. Mole (order Insectivora): teeth fine and conical for grasping insects and worms. Mule deer (order Artiodactyla): lower incisor teeth close against fleshy pad on the upper jaw to nip off vegetation which is ground between the enamel-ridged cheek teeth (premolars and molars); no canines. Beaver (order Rodentia): two pairs of chisel-like incisors used for gnawing; no canines; premolars and molars cross-ridged with enamel to grind food. Dog (order Carnivora): incisors small, canines large for stabbing, and cheek teeth differentiated for shearing and crushing.

● **Digestive System**

The mouth cavity is margined by the thin soft lips, within which are fleshy gums around the bases of the teeth. The lips and cavity are lined with soft mucous membrane. The tongue is a flexible muscular organ, attached ventrally and supported by the hyoid bones; its rough and cornified upper surface contains four kinds of papillae and microscopic taste buds. The mouth cavity is roofed by the palate; the anterior hard palate is formed of bone and crossed by fleshy ridges that help inholding food; the short fleshy soft palate behind closes the respiratory passage above during swallowing. At either side of the soft palate is the reddish tonsil, of lymphoid tissue. Four pairs of salivary glands pour watery and mucous secretions through ducts into the mouth to moisten the food; these are: the parotids, below the ears; the submaxillaries, behind the lower jaws; the sublinguals near the preceding; and the infraorbitals below the eyes.

● **Circulatory System**

The red blood corpuscles of mammals are unlike those of all lower vertebrates in being round, biconcave, and nonnucleated (oval in camels). The heart, in the thoracic cavity, is enclosed in a delicate sac of pericardium. It is completely four-chambered, as in birds, with two atria and two thick muscular ventricles. The course of blood through the heart and lungs is the same in the mammal and the bird, but in the mammal, it leaves the left ventricle of the heart through a left aortic arch. Shortly the arch gives off an innominate artery (relict of the right arch), whence the two common carotid arteries arise, and the innominate continues as the subclavian artery to the right forelimb. The arch gives off a left subclavian and then turns as the dorsal aorta, which extends posteriorly. The latter gives branches to the internal organs, body wall, and posterior limbs, then continues as the caudal artery of the tail.

● **Respiratory System**

Air enters the nostrils to pass above the palate through a maze of coiled turbinate bones covered by mucous epithelium, where it is cleaned and warmed. Behind the soft palate it crosses the pharynx to enter the glottis. This is the opening in the voice box, or larynx (Adam's apple in man), a framework of several cartilages that contain the vocal cords by which the animal's calls and squalls are produced. From the larynx the air passes down the flexible windpipe, or trachea, which is reinforced against collapse by C-shaped cartilages. The trachea continues into the thorax to divide into two bronchi. These distribute air through subdividing branches that terminate in the microscopic alveoli of the lungs. The alveoli are surrounded by pulmonary capillaries in which the oxygen–carbon dioxide exchange of external respiration occurs. The lungs are spongy elastic structures, each of three lobes. The exterior of the lungs and interior of the thorax (pleural cavity) are lined by smooth peritoneum, the pleura.

● **Excretory System**

The two kidneys lie in the lumbar region above the peritoneum. The liquid urine passes from each kidney down a duct, the ureter, to be stored in the distensible bladder that lies midventrally below the rectum. At intervals, the muscular walls of the bladder are voluntarily contracted to force urine out through the single urethra. In a female this empties at the urogenital aperture, but in a male the urethra traverses the penis.

● **Nervous System and Sense Organs**

The brain is proportionately larger than in terrestrial nonmammalian vertebrates. The olfactory lobes and brainstem are covered by the greatly enlarged cerebral hemispheres. The exterior of these is convoluted with elevations (gyri) separated by furrows (sulci); the increase of cerebral surface area is in keeping with the greater degree of intelligence displayed by mammals. The two hemispheres are joined internally by a transverse band of fibers, the corpus callosum, peculiar to mammals. The cerebellum likewise is large and conspicuously folded, being formed of a median and two lateral lobes (Fig. 17-5). Its greater development is related to the fine coordination in the animal's activities. There are 12 pairs of cranial nerves, and from the nerve cord a pair of spinal nerves passes to each body somite. The trunks of the sympathetic system lie close below the vertebrae.

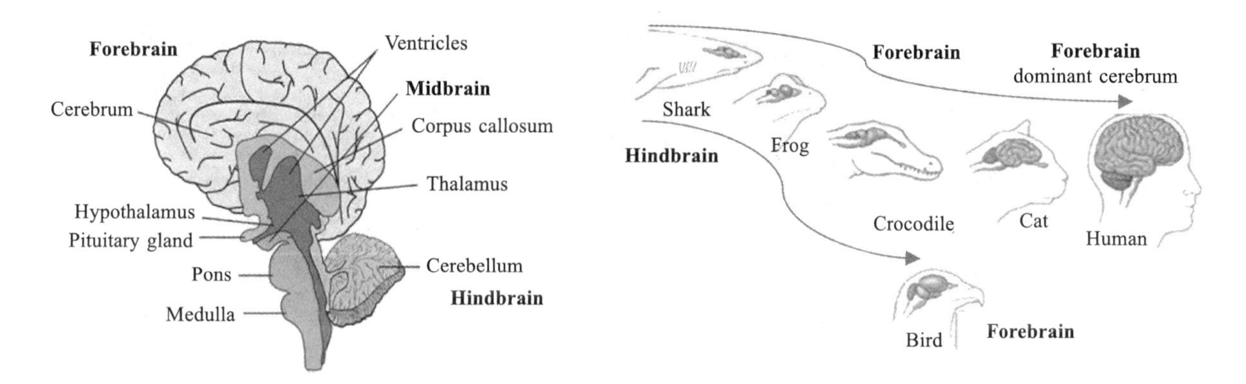

Fig. 17-5 Nervous system of mammals

The organs of taste, smell, sight, and hearing in the cat, for example, are essentially like those of man as to location and function, but the eyes are nocturnal in adaptation. Sound waves collected by the movable external ears pass into the external auditory canal that leads to the eardrum, or tympanic membrane. The middle ear of a mammal has three auditory ossicles (malleus, incus, stapes) that transfer vibrations to the inner ear—unlike the ears of birds, reptiles, and amphibians, which have only a single bone (columella) and to the cochlea, which is spirally coiled. The malleus and incus were derived by evolutionary change of function from the reptilian articular, of the lower jaw, and the quadrate, of the jaw articulation to the cranium.

● **Endocrine Glands**

In the cat, for example, endocrine glands comprise the pituitary, thyroid, parathyroids, adrenals, islets of Langerhans, and gonads. Their functions are much as in other mammals.

● **Reproductive System**

In a male, the two testes lie within the scrotum, a skin-covered double extension of the abdominal cavity suspended below the anus. From each testis the spermatozoa are gathered in a network of minute tubules in the epididymis to enter the sperm duct, or ductus deferens. This, together with blood vessels and nerves, comprises a spermatic cord that enters the abdomen through a small inguinal canal. The two ductus deferens enter the base of the urethra, which is a common urinogenital canal through the male copulatory organ, or penis, which serves to transfer sperm into the vagina of a female during copulation. Two small accessory glands, the prostate gland around the base of the urethra and the Cowper or bulbourethral glands posteriorly, provide secretions that aid in the transfer of sperm.

The structure and types of placenta are shown in Fig. 17-6.

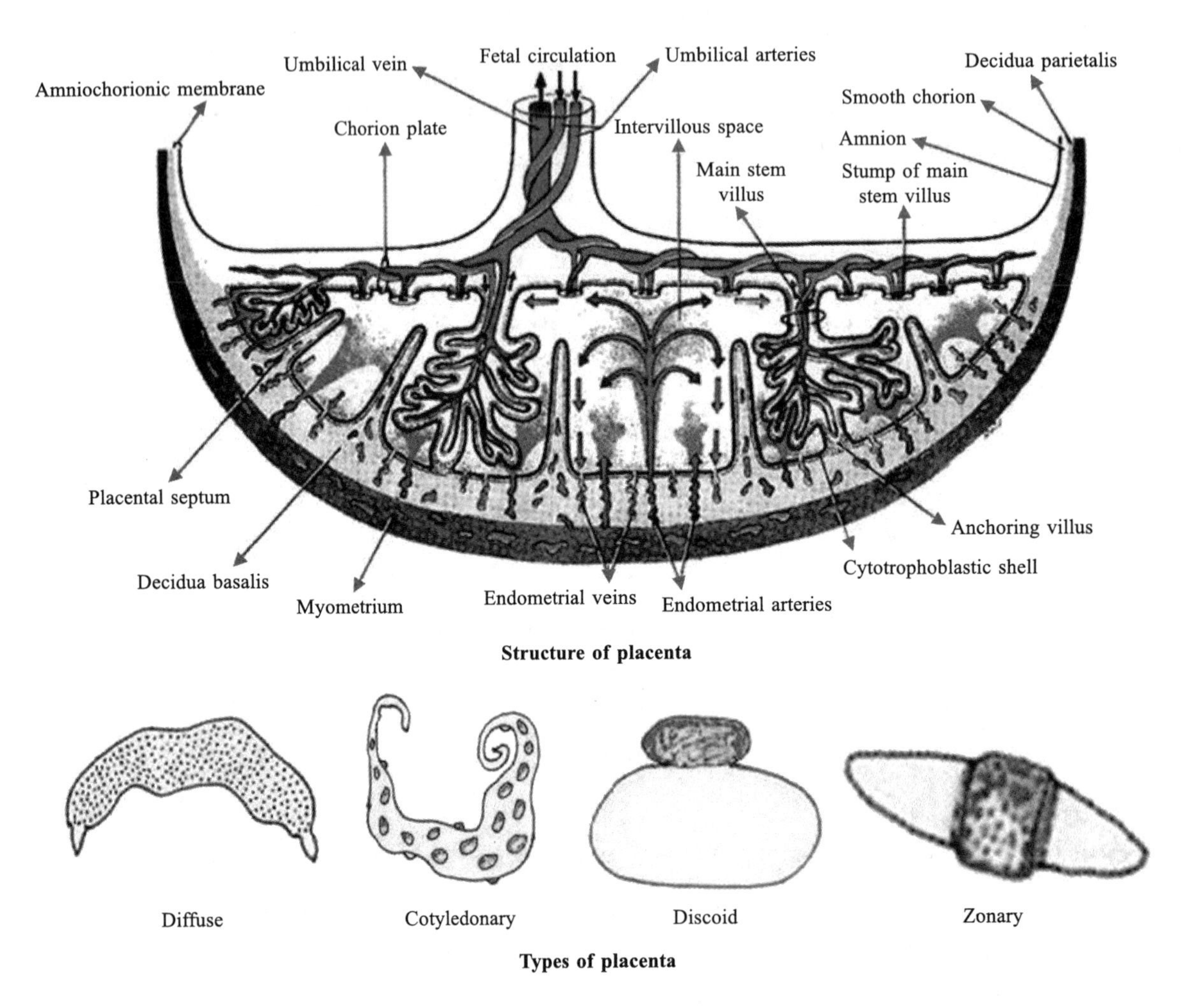

Fig. 17-6 Structure and types of placenta

17-2 Classifications of Mammals

Mammalia is the largest class in the animal kingdom. Based on their reproduction, they are classified into three subclasses: Eutheria, Metatheria, and Prototheria.

● **Eutheria**

Mammals under this subclass give birth to young ones. The young ones are developed inside the mother and derive nutrition through the placenta from the mother. Furthermore, it consists of more than 16 orders. The following orders are examples.

(1) Insectivora

Testes are abdominal. The water shrew is the tiniest mammal which is as large as a human thumb. Examples: shrews, moles and hedgehogs (Fig. 17-7).

(2) Dermoptera

A hairy skin fold called patagium extends like a parachute from neck to tail for gliding, e.g. flying lemurs. Actually, flying lemurs are neither true lemurs nor do they fly.

(3) Chiroptera

They are flying mammals. The forelimbs are modified into wings, e. g. bats and flying foxes (Fig. 17-8). The vampire bats feed on the blood of mammals including man.

(4) Edentata

They are toothless. This order includes the armadillos and sloths of South America.

(5) Pholidota

The body of these mammals is covered with overlapping horny scales with sparse hair in between. Teeth are absent, e.g. *Manis* (scaly ant eater or pangolin).

Fig. 17-7 Insectivora, Dermoptera, Edentata, and Pholidota

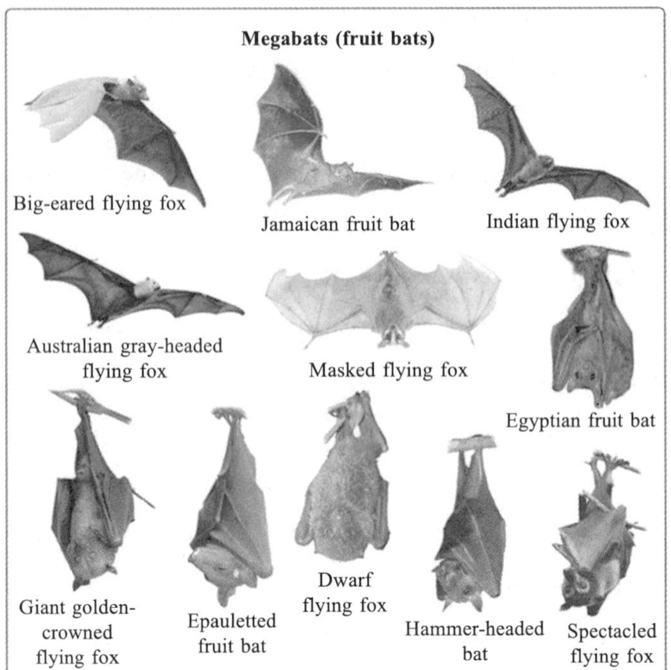

Fig. 17-8 Chiroptera

(6) Primates

Primates have highly developed brain. The living primates include prosimians (meaning before monkeys) and simians. The prosimians include lemurs, lorises and tarsiers. The simians include monkeys, apes and men (Fig. 17-9).

Fig. 17-9 Primates

(7) Rodentia

They have one pair of sharp chisel-like incisors in each jaw. The canines are absent, leaving a toothless space, the diastema in the jaw no canines, e. g. rats, mice, squirrels, guinea-pigs and porcupines (Fig. 17-10).

Fig. 17-10 Rodentia

(8) Lagomorpha

They have two pairs of incisors in the upper jaw and one pair of incisors in the lower jaw and no canines, e.g. rabbits and hares.

(9) Cetacea

They have fish-like body, well adapted for aquatic life. They have fin-like forelimbs, but no hindlimbs. Testes are abdominal. The skin has a thick layer of fat called blubber serving as reserve food, an insulator for reducing the specific gravity. Pinnae are reduced or absent. Hair is only on lips. They do not have sweat and oil glands, e. g. whales, dolphins and porpoises (Fig. 17-11). Blue whale is the largest living animal. Whales normally lack pelvic girdle and hindlimbs. The Greenland right whales, however, possess vestiges of pelvic girdles and bones of hindlimbs inside the body.

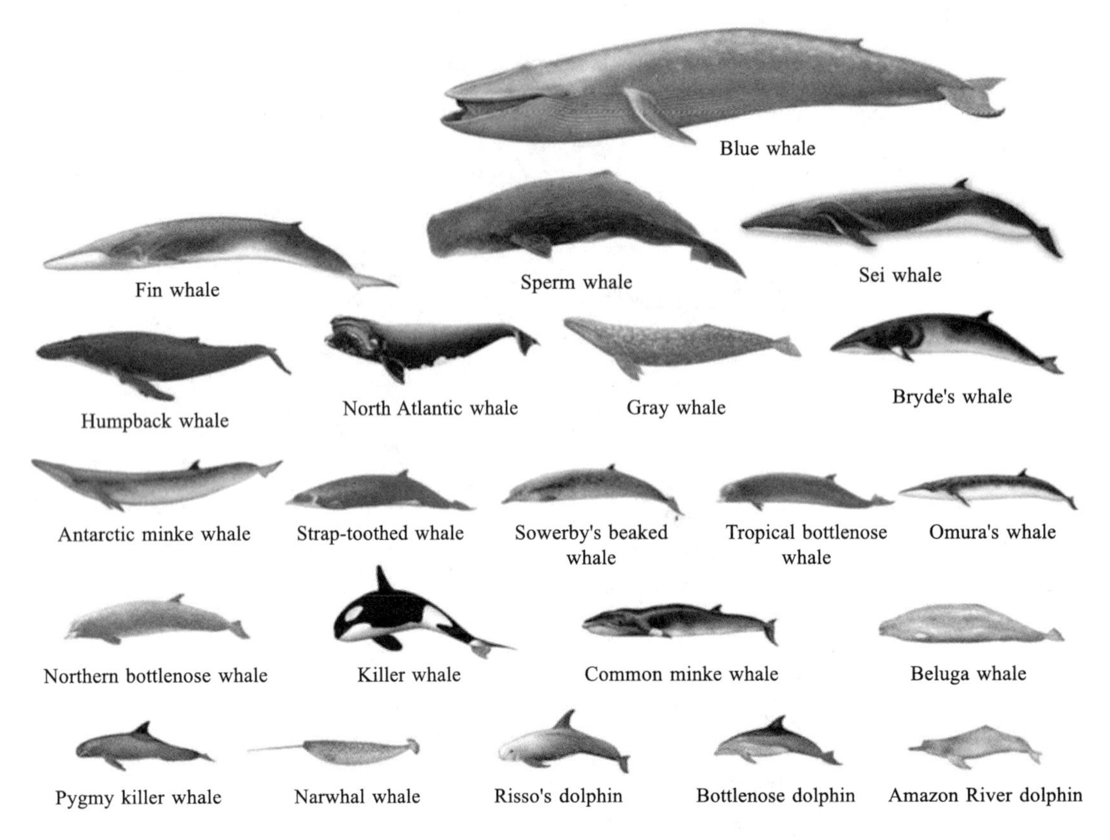

Fig. 17-11 Cetacea

(10) Carnivora

They are flesh eating mammals. These animals have sharp pointed canines, strong jaws and well-developed claws, e.g. dogs, cats, wolf, jackal, fox, cheetah, lion, tiger, hyena, mongooses, bear, pandas, otter, seal, walrus, sea lion. Cheetah is the fastest runner. It can cover a distance of 120 km in one hour (Fig. 17-12).

(11) Proboscidea

They have a long muscular trunk. They are thick skinned animals hence called pachyderms. They are the largest land animals, e.g. elephants (Fig. 17-13).

(12) Sirenia

They are herbivorous aquatic mammals with fin-like forelimbs and no hindlimbs. They have few hairs and do not have external ears. They have thick blubber. Testes are abdominal. The males have tusks, e.g. manatees (Fig. 17-14), dugongs.

(13) Perissodactyla

They are herbivorous odd-toed hoofed mammals or ungulates or hoofed which have an odd number of toes (1 or 3). True horns with a bony core are never present. The stomach is of nonruminating type (these are not cud chewing animals), e.g. horses, asses, mules, zebras, tapirs and rhinoceros (Fig. 17-15).

Fig. 17-12 Carnivora

Fig. 17-13 Elephant

Fig. 17-14 Manatee

Fig. 17-15 Perissodactyla

(14) Artiodactyla

They are herbivorous even toed hoofed mammals or ungulates which have even number of toes (2 or 4). True horns or antlers are present in many animals of this order. Many even toed hoofed mammals like cow and camel are ruminants or cud-chewing. The four chambered stomach of cow is capable of digesting cellulose of plant materials by micro-organisms present in the rumen (first part of their stomach). Examples: cows, buffaloes, sheep, goats, deer, antelopes, yaks, camels, giraffes, pigs and hippopotamuses (Fig. 17-16).

Deer　　Giraffe　　Hippopotamus

Buffalo　　Antelope

Fig. 17-16 Artiodactyla

● **Metatheria**

Now they are found mainly in Australia, New Guinea and South America. Females have a marsupium or brood-pouch for rearing young ones. Infraclass Metatheria includes one order—Marsupialia. Mammals of this order are called marsupials or pouched mammals, e. g. *Macropus*, *Didelphis* (opossum) and *Phascolarctos* (koala) (Fig. 17-17).

● **Prototheria**

Prototherians are considered to be the most primitive mammals which are only restricted in Australia and its neighbouring islands (Tasmania, New Guinea). Besides egg-laying habit, they have several reptilian characters including a cloaca. They lay eggs containing ample amount of yolk. Subclass Prototheria includes one order—Monotremata, e. g. *Ornithorhynchus* (platypus), *Tachyglossus* (echidna) (Fig. 17-18).

Kangaroo

Opossum

Koala

Thylacine

Fig. 17-17 Metatheria

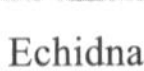

Echidna

Platypus

Fig. 17-18 Prototheria

17-3 学习要点

- 胎生、哺乳的概念及其在脊椎动物演化史上的意义。
- 哺乳类躯体结构的基本特征。
- 哺乳纲的分类；真兽亚纲主要水生种类代表种的特征及经济意义；哺乳类与人类的关系。
- 心脏及其动脉弓随动物由水生向典型陆生适应转变所发生的结构变化。
- 反刍动物胃的结构。
- 哺乳动物胎盘的类型。

17-4 巩固测验

【名词】

膈肌、胼胝体、次生腭、哺乳、胎生

【选择题】

1. （　　）是哺乳动物所特有的肌肉。

A. 咀嚼肌　　B. 皮肤肌

C. 肋间肌　　D. 膈肌

2. 下列皮肤的特征和功能，不属于哺乳动物的是（　　）。

A. 体表被毛　　B. 皮肤衍生物多

C. 皮肤干燥，腺体不发达　　D. 较厚的表皮真皮层

3. 下列不属于哺乳动物的骨骼鉴别特征的是（　　）。

A. 颈椎7枚　　B. 异型齿

C. 开放式骨盆　　D. 两个枕髁

4. 下列特征与哺乳动物消化功能增强无关的是（　　）。

A. 出现口腔消化　　B. 消化腺发达

C. 具有特殊的膈肌　　D. 草食动物发达的盲肠

5. 哺乳动物尿液中排泄的主要代谢废物是（　　）。

A. 尿素　　B. 尿酸

C. 氨基酸　　D. 肾素

【简答】

1. 胎生、哺乳在动物演化史上具有哪些重要意义？

2. 比较原兽亚纲、后兽亚纲和真兽亚纲的主要特征。

Chapter 18 Animal Evolution

18-1 Summary

Animal evolution is the theory that modern animals are the modified descendants of animals that formerly existed and that these earlier forms descended from still earlier and different organisms. Animals are multicellular organisms that feed by ingestion of other organisms or their products, being unable to derive energy through photosynthesis or chemosynthesis. Animals are currently classed into about 30 to 35 phyla, each of which has evolved a distinctive body plan or architecture.

All phyla began as invertebrates, but lineages of the phylum Chordata developed the internal skeletal armature, with spinal column, which was exploited in numerous fish groups and which eventually gave rise to terrestrial vertebrates. The number of phyla is uncertain partly because most of the branching patterns and the ancestral body plans from which putative phyla have arisen are not yet known. For example, arthropods may have all diversified from a common ancestor that was a primitive arthropod, in which case they may be grouped into a single phylum; or several arthropod groups may have evolved independently from non-arthropod ancestors, in which case each such group must be considered a separate phylum. So far as known, all animal phyla began in the sea.

Some features of the cells of primitive animals resemble those of the single-celled Protozoa, especially the flagellates, which have long been believed to be animal ancestors. Molecular phylogenies have supported this idea and also suggest that the phylum Coelenterata arose separately from all other phyla that have been studied by this technique. Thus, animals may have evolved at least twice from organisms that are not themselves animals, and represent a grade of evolution and not a single branch (clade) of the tree of life. Sponges have also been suspected of an independent origin, and it is possible that some of the extinct fossil phyla arose independently or branched from sponges or cnidarians (Fig. 18-1).

The earliest undoubted animal fossils (the Ediacaran fauna) are soft-bodied, and first appear in marine sediments nearly 650 million years old. This fauna lasted about 50 million years and consisted chiefly of cnidarians or cnidarian-grade forms, though it contains a few enigmatic fossils that may represent groups that gave rise to more advanced phyla. Then, nearly 570 million years ago, just before and during earliest Cambrian time, a diversification of body architecture began that produced most of the living phyla as well as many extinct groups. The body plans of some of these groups involved mineralized skeletons which, as these are more easily preserved than soft tissues, created for the first time an extensive fossil record. The soft-bodied groups were markedly diversified, though their record is so spotty that their history cannot be traced in detail. A single exceptionally preserved soft-bodied fauna from the Burgess Shale of British Columbia that is about 530 million years old contains not only living soft-bodied worm phyla, but extinct groups that cannot be placed in living phyla and do not seem to be ancestral to them.

Following the early phase of rampant diversification and of some concurrent extinction of phyla and their major branches, the subsequent history of the durably skeletonized groups can be followed in a general way in the marine fossil record. The composition of the fauna changed continually, but three major associations can be seen: one dominated by the arthropod like trilobites during the early Paleozoic, one dominated by articulate brachiopods and crinoids (Echinodermata) in the remaining Paleozoic, and one dominated by gastropod (snail) and bivalve (clam) mollusks during the Mesozoic and Cenozoic. The mass extinction at the close of the Paleozoic that caused the contractions in so many groups may have extirpated over 90% of marine species and led to a reorganization of marine community structure and composition into a modern mode. Resistance to this and other extinctions seems to have been a major factor in the rise of successive groups to dominance. Annelids, arthropods, and mollusks are the more important invertebrate groups that made the transition to land. The outstanding feature of terrestrial fauna is the importance of the insects, which appeared in the late Paleozoic and later radiated to produce the several million living species, surpassing all other life forms combined in this respect.

The phylum Chordata consists largely of animals with a backbone, the Vertebrata, including humans. The group, however, includes some primitive invertebrates, the protochordates: lancelets, tunicates, acorn worms, pterobranchs, and possibly the extinct graptolites and conodonts. The interrelationships of these forms are not well understood. With the exception of the colonial graptolites, they are soft-bodied and have only a very limited fossil record. They suggest possible links to the Echinodermata in developmental, biochemical, and morphological features. In addition, some early Paleozoic fossils, the carpoids, have been classified alternatively as chordates and as echinoderms, again suggesting a link. In spite of these various leads, the origin of the chordates remains basically unclear.

Chordates are characterized by a hollow, dorsal, axial nerve chord, a ventral heart, a system of slits in the larynx that serves variously the functions of feeding and respiration, a postanal swimming tail, and a notochord that is an elongate supporting structure lying immediately below the nerve chord. The protochordates were segmented, although sessile forms such as the tunicates show this only in the swimming larval phase.

The first vertebrates were fishlike animals in which the pharyngeal slits formed a series of pouches that functioned as respiratory gills. An anterior specialized mouth permitted ingestion of food items large in comparison with those of the filter-feeding protochordates. Vertebrates are first known from bone fragments found in rocks of Cambrian age, but more complete remains have come from the Middle Ordovician. Innovations, related to greater musculoskeletal activity, included the origin of a supporting skeleton of cartilage and bone, a larger brain, and three pairs of cranial sense organs (nose, eyes, and ears). At first, the osseous skeleton served as protective scales in the skin, as a supplement to the notochord, and as a casing around the brain. In later vertebrates, the adult notochord is largely or wholly replaced by bone, which encloses the nerve chord to form a true backbone. All vertebrates have a heart which pumps blood through capillaries, where exchanges of gases with the external media take place. The blood contains hemoglobin in special cells which carry oxygen and carbon dioxide. In most fishes, the blood passes from the heart to the gills and thence to the brain and other parts of the body. In most tetrapods, and in some fishes, blood passes to the lungs, is returned to the heart after oxygenation, and is then pumped to the various parts of the body.

The jawless fishes, known as Agnatha, have a sucking-rasping mouth apparatus rather than true jaws. They enjoyed great success from the Late Cambrian until the end of the Devonian. Most were heavily armored, although a few naked forms are known. They were weak swimmers and lived mostly on the bottom. The modern parasitic lampreys and deep-sea scavenging hagfish are the only surviving descendants of these early fish radiations.

In the Middle to Late Silurian arose a new type of vertebrate, the Gnathostomata, characterized by true jaws and teeth. They constitute the great majority of fishes and all tetrapod vertebrates. The jaws are modified elements of the front parts of the gill apparatus, and the teeth are modified bony scales from the skin of the mouth. With the development of jaws, a whole new set of ecological opportunities was open to the vertebrates. Along with this, new swimming patterns appeared, made possible by the origin of paired fins, forerunners of which occur in some agnathans.

Four groups of fishes quickly diversified. Of these, the Placodermi and Acanthodii are extinct. The Placodermi were heavily armored fishes, the dominant marine carnivores of the Silurian and Devonian. The Acanthodii were filter-feeders mostly of small size. They are possibly related to the dominant groups of modern fishes, the largely cartilaginous Chondrichthyes (including sharks, rays, and chimaeras) and the Osteichthyes (the higher bony fishes). These also arose in the Late Silurian but diversified later.

The first land vertebrates, the Amphibia, appeared in the Late Devonian and were derived from an early group of osteichthyans called lobe-finned fishes, of which two kinds survive today, the Dipnoi or lungfishes, and the crossopterygian coelacanth *Latimeria*. They were lung-breathing fishes that lived in shallow marine waters and in swamps and marshes. The first amphibians fed and reproduced in or near the water. True land vertebrates, Reptilia, with a modified (amniote) egg that could survive on land, probably arose in the Mississippian.

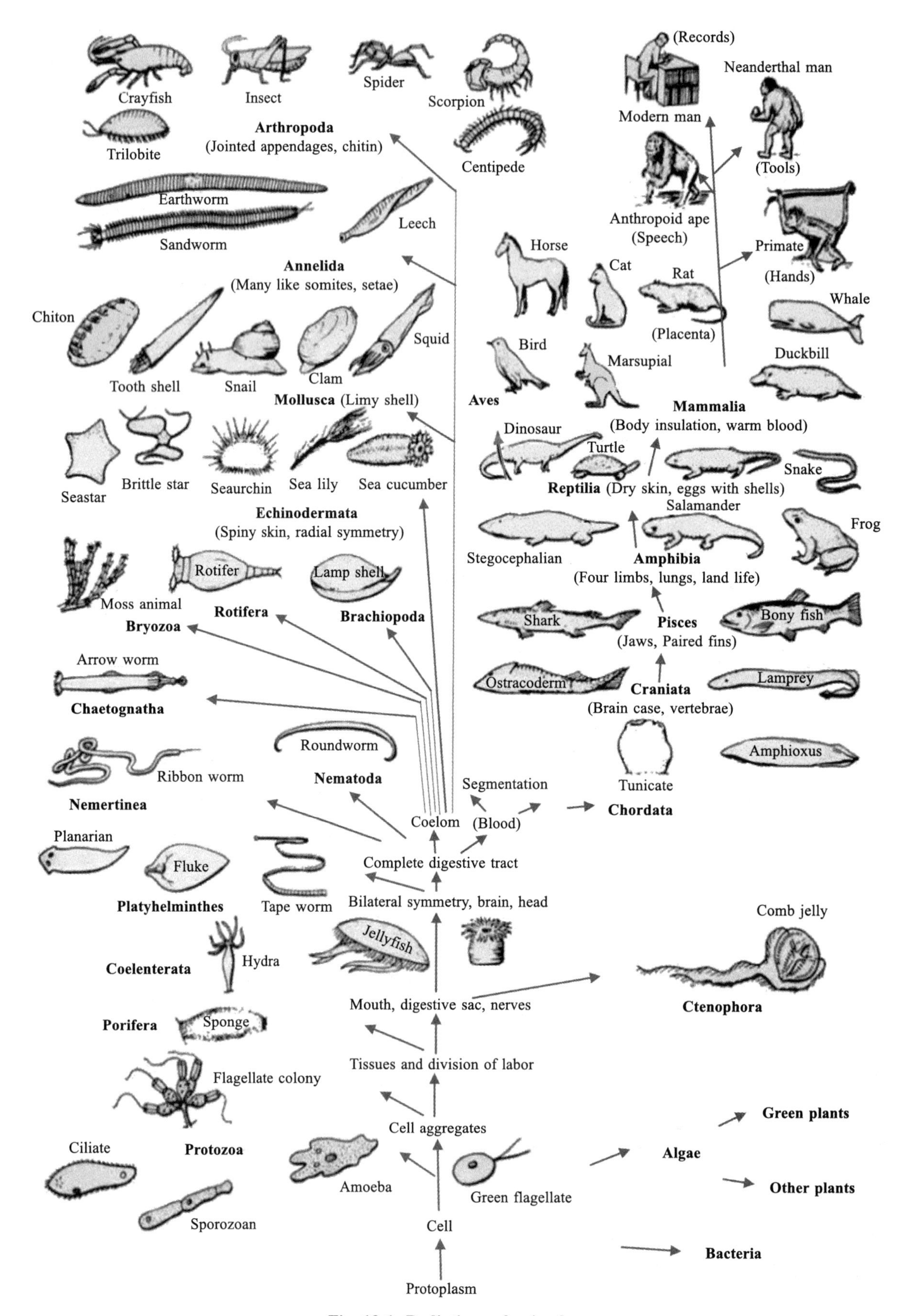

Fig. 18-1 Radiations of animals

By the Middle Pennsylvanian, a massive radiation of reptiles was in process. The most prominent reptiles belong in the Diapsida: dinosaurs, lizards and snakes, and pterosaurs (flying reptiles). The birds, Aves, which diverged from the dinosaur radiation in the Late Triassic or Early Jurassic, are considered to be feathered dinosaurs, and thus members of the Diapsida, whereas older authorities prefer to treat them as a separate case. In addition, there were several Mesozoic radiations of marine reptiles such as ichthyosaurs and plesiosaurs. Turtles (Chelonia) first appeared in the Triassic and have been highly successful ever since.

The line leading to mammals can be traced to primitive Pennsylvanian reptiles, Synapsida, which diversified and spread worldwide during the Permian and Triassic. The first true mammals, based on characteristics of jaw, tooth, and ear structure, arose in the Late Triassic. Derived mammals, marsupials (Metatheria) and placentals (Eutheria), are known from the Late Cretaceous, but mammalian radiations began only in the early Cenozoic. By the end of the Eocene, all the major lines of modern mammals had become established. Molecular analyses (blood proteins, deoxyribonucleic acid, ribonucleic acid) of living mammals show that the most primitive group of placentals is the edentates (sloths, armadillos, and anteaters). An early large radiation included the rodents, primates (including monkeys, apes, and humans), and bats, possibly all closely related to the insectivores and carnivores. The newest radiations of mammals are of elephants and sea cows, while the whales are related to the artiodactyls (cattle, camels).

18–2 The Origin of Animal Multicellularity

In particular, multicellularity is defined as a condition or state of having or being comprised of many cells, each seemingly performing distinct function(s). One of the popular theories held is the "Gastraea Theory of Haeckel". Accordingly, multicellularity first occurred when cells of the same species group together in a blastula-like colony (Fig. 18-2). In due time, certain cells in the colony underwent cell differentiation. Still, this theory seems inadequate to explain the origin of multicellularity.

What scientists agree on is that, in essence, multicellularity occurred several times in biological history. In Neoproterozoic Era, particularly in the Ediacaran Period (around 600 million years ago), the first multicellular form emerged. Also in this period, sponge-like organisms evolved based on the recovered fossils of Ediacaran biota. Correspondingly, they were presumed to be the first animals. They resemble the sponges (choanocytes) with size ranging from 1 cm to less than 1 m. The earliest ancestors of multicelled animals could likely be sponge-like because a sponge has no organs but an assemblage of different cells with specialized functions.

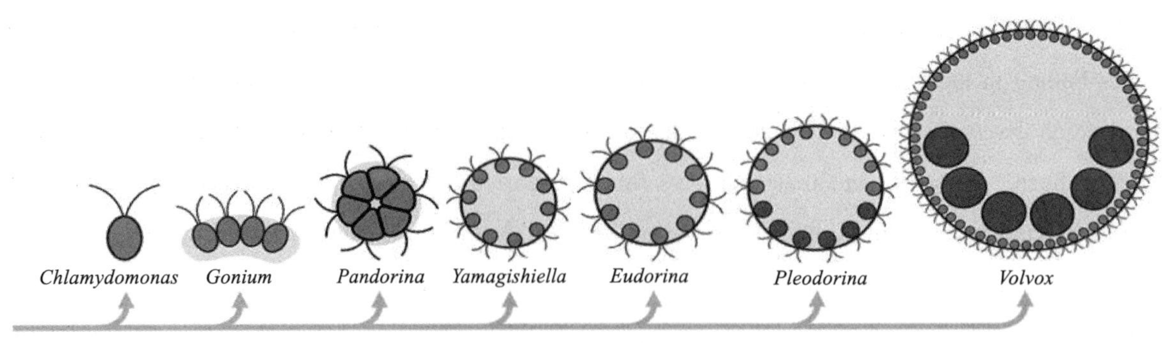

Fig. 18-2 Multicellularity of chlorophytes

18-3 Embryonic Development

● **Fertilization**

Fertilization is the process beginning with penetration of the secondary oocyte by the sperm and completed by fusion of the male and female pronuclei. If one sperm does make its way into the fallopian tube and burrows into the egg, it fertilizes the egg. The egg changes so that no other sperm can get in.

● **Cleavage**

Cleavage can take place in two ways: holoblastic (total) cleavage or meroblastic (partial) cleavage. The type of cleavage depends on the amount of yolk in the eggs. In placental mammals (including humans) where nourishment is provided by the mother's body, the eggs have a very small amount of yolk and undergo holoblastic cleavage. Other species, such as birds, with a lot of yolk in the egg to nourish the embryo during development, undergo meroblastic cleavage. Patterns of cleavage depends upon amount of yolk and distribution of yolk. These are following types (Fig. 18-3).

(1) Radial Cleavage Pattern

In radial cleavage, the cleavage furrow cut straight through the egg at right angle to another. It results in blastomeres appearing to be arranged radially. Examples: sponge, Coelenterata.

(2) Biradial Cleavage Pattern

The cleavage furrow of first three division do not cut straight the axis of the egg. It determines by radial symmetry in animal. Example: Ctenophora.

(3) Spiral Cleavage Pattern

The mitotic spindle of the third cleavage in the four blastomeres are laid down obliquely and are arranged in a sort spiral. It results in four blastomeres of upper tier, which do not lie over the corresponding blastomeres of lower tier. Examples: annelids, molluscs, nemerteans.

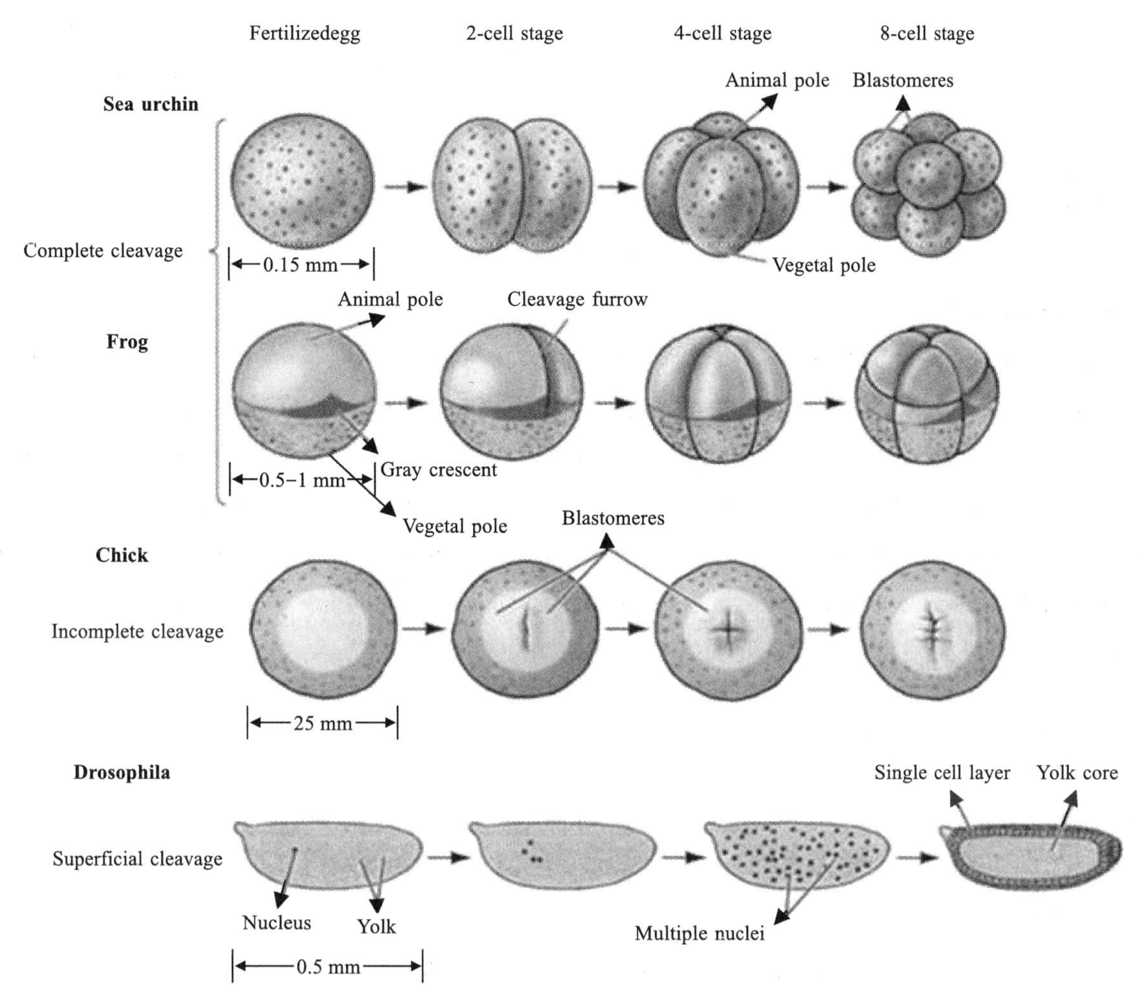

Fig. 18-3 Patterns of cleavage

Spiral cleavage is of two types:

Dextral or dextrotrophic: The blastomere rotates clockwise (i. e. right hand spiral). Example: Mollusca.

Sinistral or levotrophic: The blastomere rotates anticlockwise (i.e. left hand coiling). Examples: flat worm, round worm, annelids.

(4) Bilateral Cleavage Pattern

Two of the four blastomeres are smaller than the other two and establish bilateral symmetry at the four-cell stage. Examples: Echinodermata, Chordata.

● **Blastula**

A single-celled zygote will undergo multiple rounds of cleavage, or cell division, in order to produce a ball of cells, called a blastula, with a fluid-filled cavity in its center, called a blastocoel (Fig. 18-4).

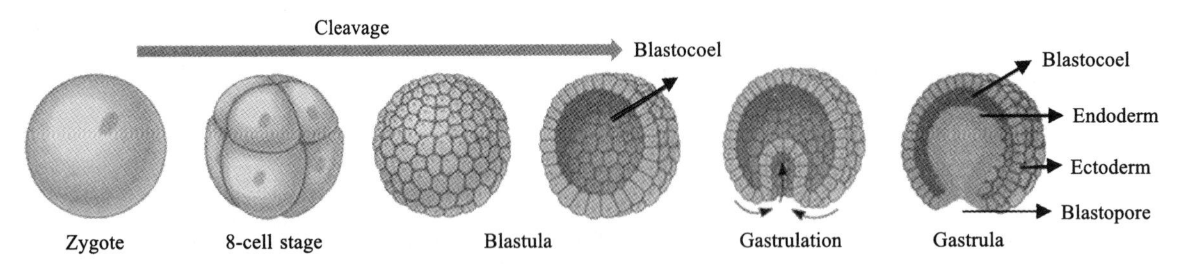

Fig. 18-4 Zygote, blastla and gastrula

● Gastrulation

The typical blastula is a ball of cells. The next stage in embryonic development is the formation of the body plan. The cells in the blastula rearrange themselves spatially to form three layers of cells in a process known as gastrulation. During gastrulation, the blastula folds upon itself to form the three layers of cells. Each of these layers is called a germ layer, which differentiate into different organ systems. Although gastrulation patterns exhibit enormous variation throughout the animal kingdom, they are unified by the five basic types of cell movements that occur during gastrulation: invagination, involution, ingression, delamination, epiboly (Fig. 18-5).

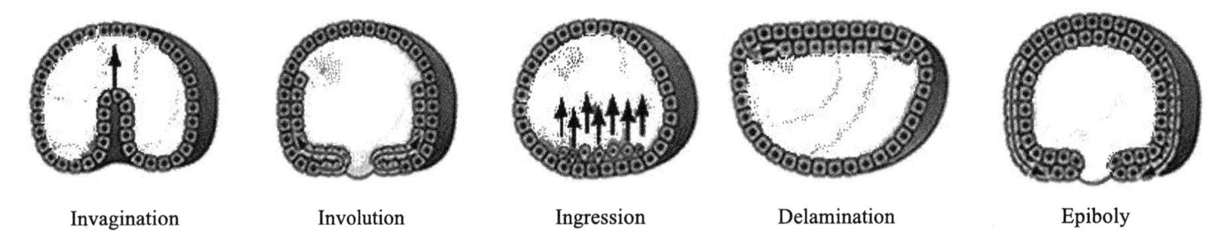

Fig. 18-5 Basic types of cell movements during gastrulation

● Mesoderm and Coelom

The mesoderm is a germ layer present in animal embryos that will give rise to specialized tissue types. The mesoderm is one of three germ layers found in triploblastic organisms; it is found between the ectoderm and endoderm. All bilaterally-symmetrical animals are triploblasts, whereas some simpler animals such as cnidarians and ctenophores (jellyfish and comb jellies) have only two germ layers and are diploblasts. The mesoderm plays an important role in animal development. It goes on to form many central structures including the skeletal system, muscular system, and the notochord.

The coelom is a body cavity found in metazoans (animals that develop from an embryo with three tissue layers: ectoderm, mesoderm, and endoderm). The cells in each tissue layer become differentiated during development, becoming different tissues, organs, and a digestive tract. Derived from the mesoderm, the coelom is found between the intestinal canal and the body wall, lined with mesodermal epithelium. The mesodermal tissue also goes on to form the blood, bones, digestive tract, gonads, kidneys, and other organs. Organisms that possess a true coelom are called (true) coelomates.

True coelomates are often grouped into two categories: protostomes and deuterostomes. This distinction is based on patterns of cell division, coelom formation (Fig. 18-6), and the fate of the blastopore. In protostomes, the blastopore becomes the mouth. In deuterostomes, the blastopore becomes the anus. Organisms that possess a body cavity that is not fully lined with mesodermal epithelium are called pseudocoelomates, while organisms that lack a body cavity are called acoelomates.

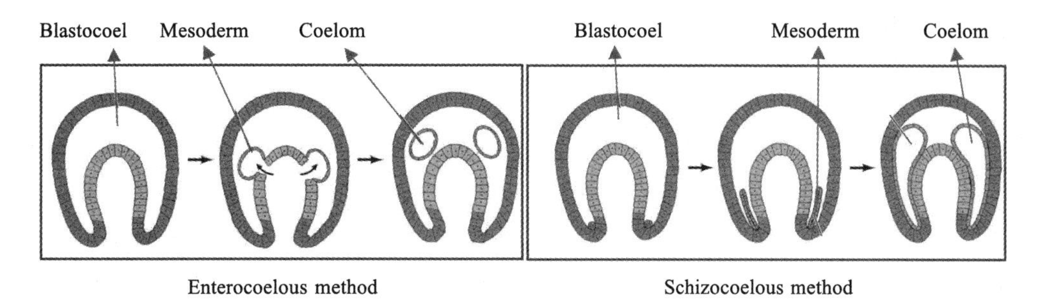

Fig. 18-6 Methods of coelom formation

18–4 Biogenetic Law

Biogenetic law states that the earlier stages of embryos of species advanced in the evolutionary process, such as humans, resemble the embryos of ancestral species, such as fish (Fig. 18-7). The law refers only to embryonic development and not to adult stages; as development proceeds, the embryos of

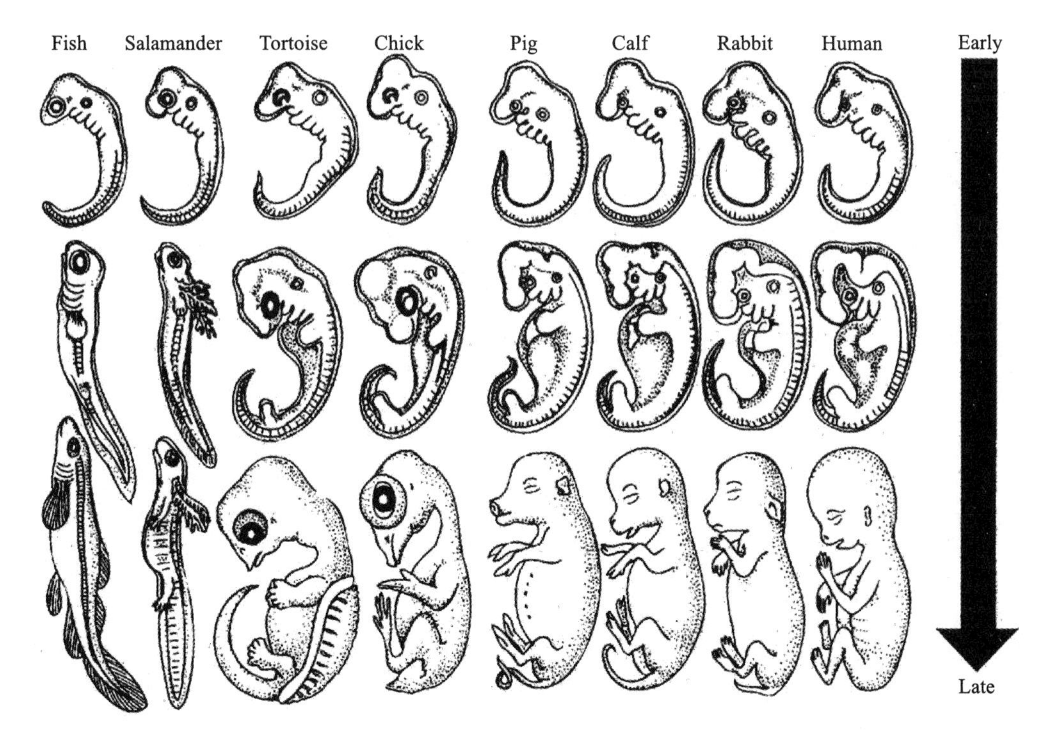

Top: All are much alike in the earliest stage. Middle: Differentiation is evident, but the four mammals are quite small. Bottom: Later the distinctive characteristics of each become evident.

Fig. 18-7 Haeckel's comparison of early embryonic stages across vertebrate groups (Haeckel, 1891)

different species become more and more dissimilar. An early form of the law was devised by the 19th-century Estonian zoologist, who observed that embryos resemble the embryos, but not the adults, of other species. A later, but incorrect, theory of the 19th-century German zoologist states that the embryonic development (ontogeny) of an animal recapitulates the evolutionary development of the animal's ancestors (phylogeny).

18-5 Skeletal Systems

The skeleton may be a shell or other external covering (exoskeleton), as on corals, mollusks, and arthropods, or internal (endoskeleton), as with vertebrates (Fig. 18-8–Fig. 18-10). It is rigid on corals, many mollusks, and others, but variously jointed and movable in echinoderms, arthropods, and vertebrates. Exoskeletons serving as defensive armor were present on fossil animals such as the trilobites, primitive fishlike ostracoderms, early amphibians (labyrinthodonts), and some ancient reptiles (dinosaurs); they occur also on living brachiopods, most mollusks, barnacles, some fishes, the turtles and tortoises, and the armadillo. An exoskeleton limits the ultimate size of an animal and may become so heavy that the organism must remain fixed. This is because the internal muscles cannot be large and powerful enough to move the heavy framework. The internal skeleton of a vertebrate involves far less limitation, and some vertebrates have attained huge size. These include the brontosaurs and other fossil reptiles and the living elephants and rhinoceroses.

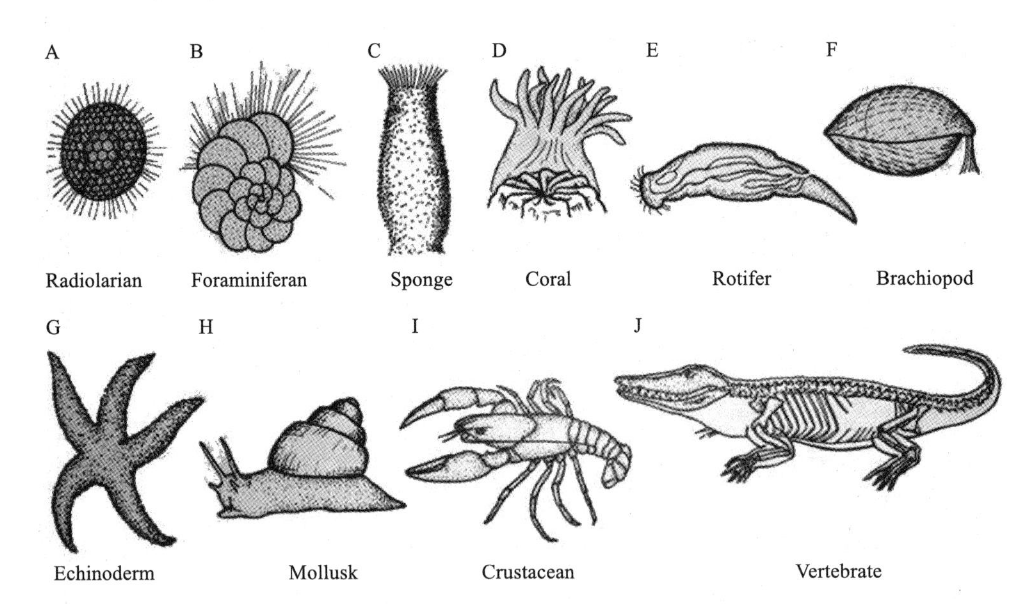

A. Radiolarian, framework of strontium sulfate. B. Foraminiferan, limy shell. C. Sponge, many minute limy or glassy spicules. D. Coral, solid calcareous (limy) cup with partitions. E. Rotifer, firm "glassy" cuticle. F. Brachiopod, two limy shells. G. Echinoderm, internal jointed skeleton of limy plates. H. Mollusk, limy shell. I. Crustacean, complete exoskeleton with chitin. J. Vertebrate, skull, vertebrae, limb girdles, and limb skeleton of bone.

Fig. 18-8 Skeletons in animals

(Storer et al., 1972)

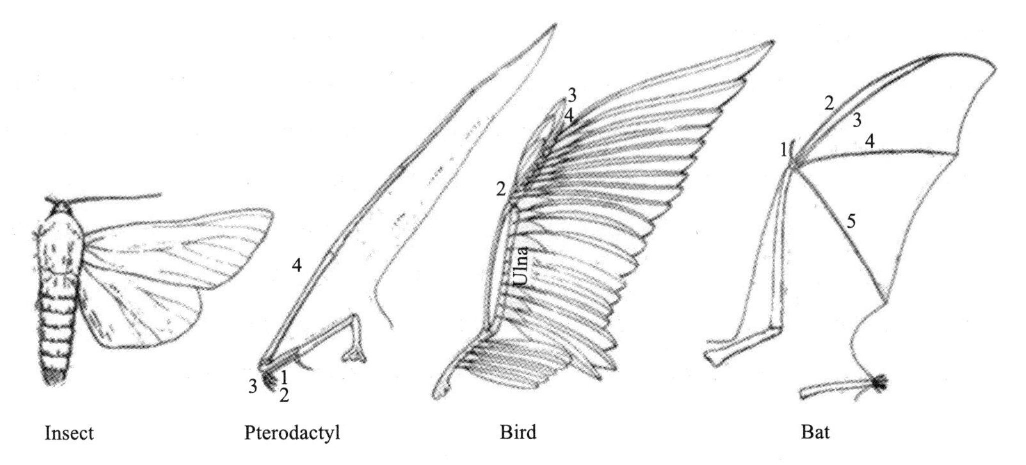

Homology in the wing bones of vertebrates, all derived from the common pattern of the forelimb in land vertebrates, but variously modified. Pterodactyl (extinct reptile) with very long fourth finger; bird with first and fifth lacking, third and fourth partly fused; bat with second to fifth fingers long.

Fig. 18-9 Analogy between wings of insects (no internal skeleton) and of vertebrates (with skeleton)—of like function but different origins

(Storer et al., 1972)

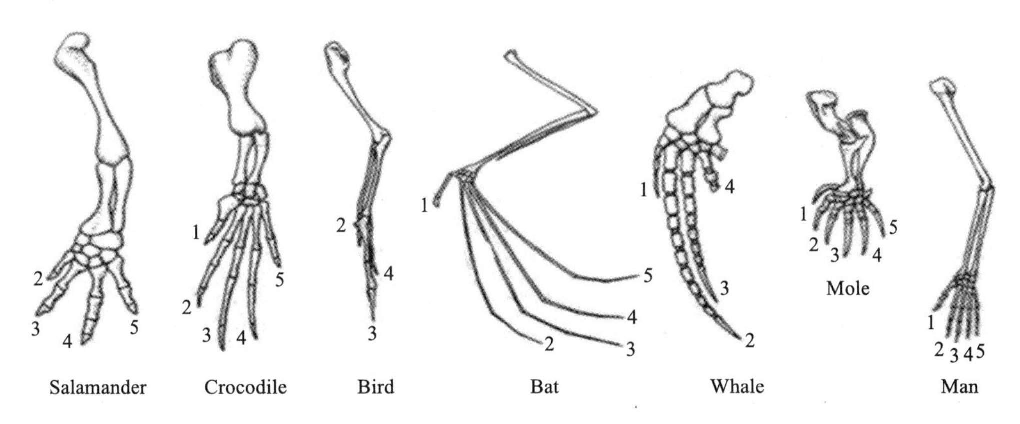

The limbs arehomologous in being composed of comparable bones (humerus, carpals, etc.) which in each kind of animal are adapted for special uses by differences in the length, shape, and bulk of the various bones, one to five digits, or fingers.

Fig. 18-10 Homology and adaptation in bones of the left forelimb in land vertebrates

(Storer et al., 1972)

18-6 Digestive Systems

Animals differ widely in their food habits. Some insects feed on the tissues or juices of a single species of plant or the blood of one kind of animal, but most animals take several or many kinds of foods. Cattle, deer, rodents, and insects that eat leaves and stems of plants are herbivorous; cats, sharks, flesh flies, and many marine animals whose food consists entirely or largely of other animals are carnivorous; and man, bears, rats, and others that eat various plant and animal materials are general feeders, or omnivorous.

Paramecia and certain other protozoans, some sea anemones, certain fishes, and tadpoles that feed on small particles, living or dead, such as plankton, are termed microphagous feeders. In contrast, most higher animals, including man, that use larger materials are macrophagous feeders. A few feed on fluids, like the mosquitoes and ticks that suck blood and the aphids that pump in plant juices. The digestive mechanism in various animals differs in general form, structural and physiologic processes according to the nature of the food manner of life, and other factors. All means for taking and using food are essentially alike in those materials from the external environment are brought into intimate contact with internal membranous surfaces where digestion and absorption can take place (Fig. 18-11).

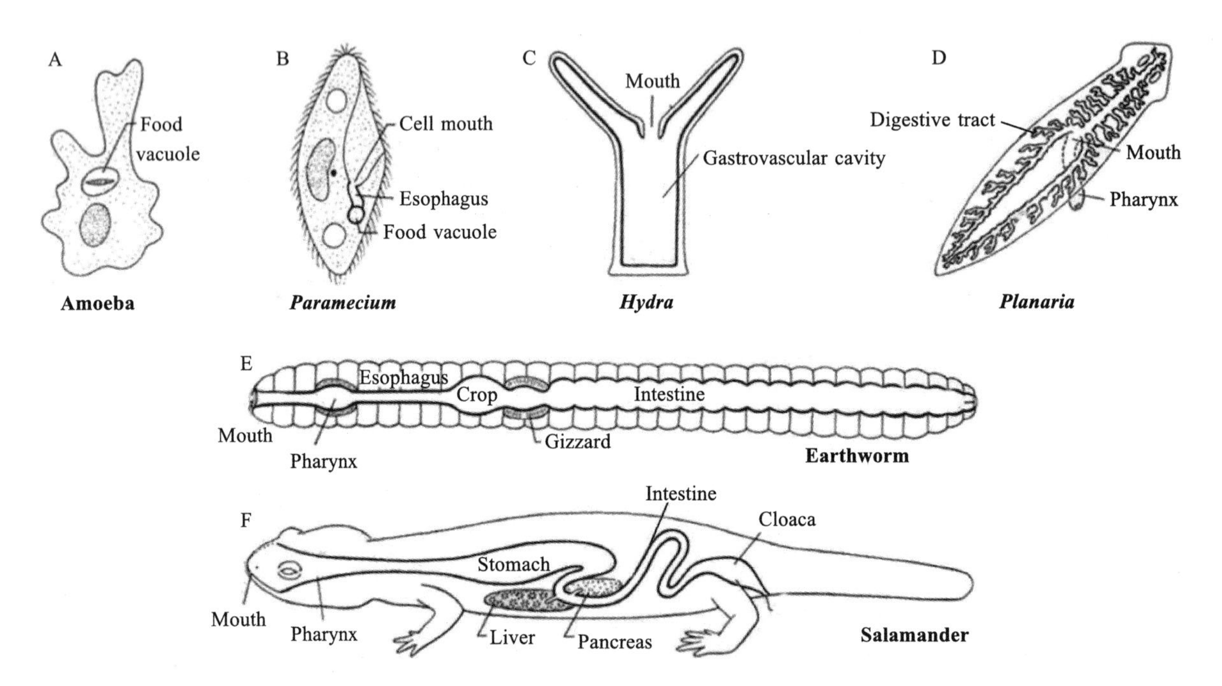

A. Amoeba, food enters at any place on cell surface. B. *Paramecium*, with definite cell mouth. C. *Hydra*, mouth and sac-like digestive cavity. D. *Planaria*, mouth and branched digestive tract, but no anus. E. Earthworm, tubular digestive tract having specialized sections, complete with terminal mouth and anus. F. Vertebrate, complete and partly coiled tract with specialized parts and digestive glands, anus at base of tail.

Fig. 18-11 Types of digestive systems in animals; diagrammatic
(Storer et al., 1972)

18–7 Circulatory Systems

Sponges, coelenterates, and flatworms have no circulatory system. Simple fusion serves to carry digested food, respiratory gases, and wastes between various parts of their bodies. In nematodes, rotifers, and entoprocts, body fluid is circulated in a pseudocoel. Most echinoderms have the coelomic fluid, hemal fluid and that in intercellular spaces variously interconnected, but the ambulacral fluid is separate. Nemerteans have one dorsal and two lateral blood vessels with many cross connections.

Pulsations in the walls of these serve to circulate the blood, which, surprisingly, contains red blood cells resembling those of vertebrates.

The blood of most invertebrates has relatively few free cells in the plasma as compared with vertebrates. Usually there are amoeboid corpuscles resembling white blood cells, some phagocytic and some aiding in transport of food or other substances. In insects, many of the cells cling to organs and become common in the plasma only after bodily injury or during molt. If a respiratory pigment is present to carry oxygen, it often is dissolved in the plasma.

The heart of invertebrates is always dorsal to the digestive tract. In mollusks it is short, lies within a thin pericardial sac, and consists of one or two thin-walled auricles that receive blood from the body and deliver it to a single muscular-walled ventricle. The latter contracts to force the blood through the vessels, or arteries, which distribute to various organs. Insects and many other arthropods have the heart as a slender dorsal tube with segmentally placed lateral openings (ostia) that receive blood from the body spaces (hemocoel) and pump it through a median aorta whence it flows through body spaces to organs and tissues. Many insects have accessory hearts which serve as boosters to propel blood through the antennae, wings, and legs.

The earthworm has a closed system with several lengthwise vessels through the body and paired transverse connecting vessels in most body segments. The circulation is produced by contractions of the middorsal vessel and by five pairs of lateral hearts in anterior body segments. In tunicates, the heart is a valveless tube lying in the pericardium, and the path of blood flow reverses at short intervals.

The hearts of vertebrates and mollusks are myogenic, the beat originating in the heart muscle itself, whereas the hearts of crustaceans and limulus are neurogenic, the beat originating with nerve ganglion cells (Fig. 18-12–Fig. 18-14).

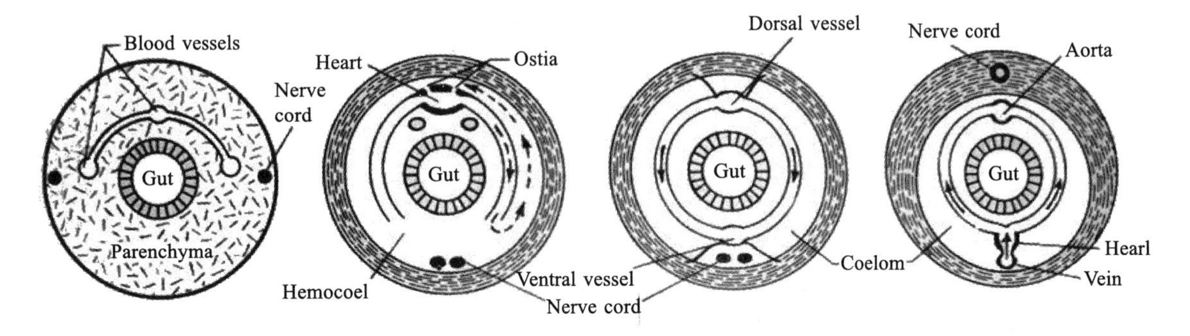

Arrows show path of blood flow. A. Nemertean. Mere system of vessels, no heart, body filled with packing tissue (parenchyma). B. Arthropod or mollusk. An open system. Dorsal heart (and aortas) from which blood escapes into body spaces—the hemocoel—and returns to the heart (some mollusks collect blood from the hemocoel in veins). Coelom reduced to small cavities (in pericardial sac, gonads, etc). C. Annelid. Closed system of vessels containing the blood; body spaces (between gut and body wall) a true coelom. D. Vertebrate. Closed system, heart ventral; a true coelom.

Fig. 18-12 Diagrammatic cross sections to show relations of circulatory system to body spaces and organs (Storer et al., 1972)

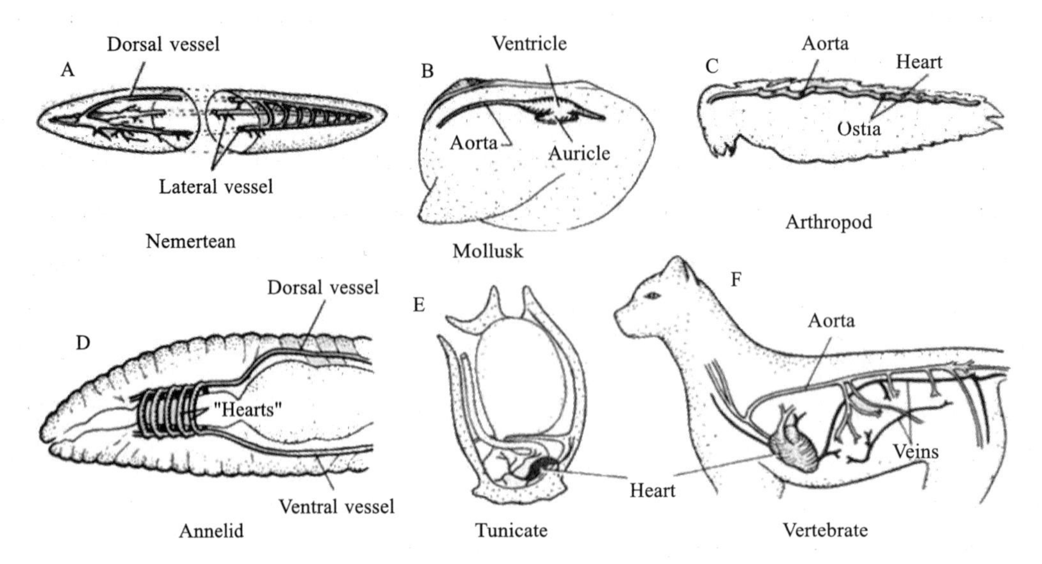

A. Nemertean (ribbon worm), simple lengthwise dorsal and lateral vessels with cross connectives. B. Mollusk (bivalve), dorsal heart with auricle (1 or 2) and ventricle, anterior, and posterior aortas; blood returns through body spaces (hemocoel)—open system. C. Arthropod (insect), dorsal tubular heart and aorta; blood returns through body spaces (hemocoel)—open system. D. Annelid (earthworm), dorsal and ventral vessels (and others) with cross connectives—closed system. E. Tunicate (sea squirt), a heart and aortas, vessels obscure; blood flow reverses. F. Vertebrate (mammal), chambered heart, definite aorta, arteries, and veins, with connections to respiratory organs—closed system.

Fig. 18-13 Types of circulatory systems in animals
(Storer et al., 1972)

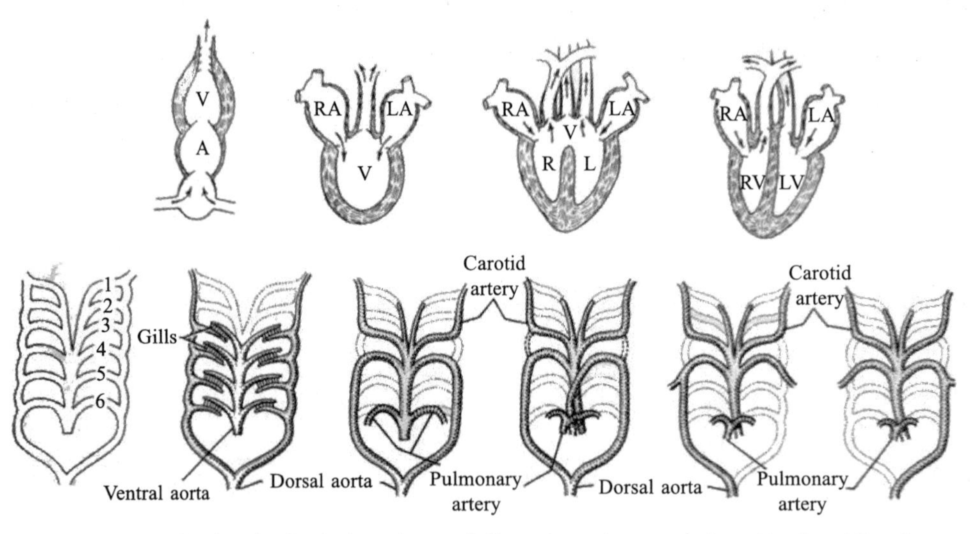

Below six pairs of arches develop in the embryos of all vertebrates but parts indicated by dotted lines later disappear. In land vertebrates, the third pair always forms the carotid arteries; the fourth becomes the systemic arches to the dorsal aorta, but only the right persists in birds and the left in mammals; the sixth arch always forms the pulmonary arteries. Above. The embryonic heart always begins with one atrium (A) and one ventricle (D); it remains thus in fishes. The atrium becomes divided (RA, LA) in amphibia; the ventricle becomes partly divided in reptiles and completely so (RV, LW) in birds and mammals. In embryos of higher forms, the arches and chambers develop progressively through the succession of stages shown. Arrows indicate paths of blood flow.

Fig. 18-14 Homology and embryonic sequence in the aortic arches and heart chambers of vertebrates
(Storer et al., 1972)

18-8 Respiratory and Excretory Systems

Ordinary respiration in different animals is performed by various respiratory organs or systems, such as the body covering, gills, lungs, or tracheae (Fig. 18-15). These structures are unlike in appearance but fundamentally the same in function; each comprises a moist permeable membrane through which molecules of oxygen and carbon dioxide diffuse readily. In accordance with the laws of gases, each gas acts independently of others. When a difference in diffusion pressure exists on the two sides of a membrane, more molecules pass toward the region of lesser pressure than in the opposite direction. The partial pressure of oxygen in the air or water is greater than that within an animal body, where it is constantly being used up, so that oxygen tends to enter any suitable membrane surface. The partial pressure of carbon dioxide is greater within the animal, so that it tends to pass outward. These exchanges occur simultaneously. In many small animals, the exchange of gases is direct, from air or water through membranes to tissue cells; but it is more complex in larger species and those with dry or nonpermeable exteriors. In the latter, respiration consists of two stages: external respiration, the exchange between environment and the respiratory organs, and internal respiration, the exchange between the body fluids and the tissue cells.

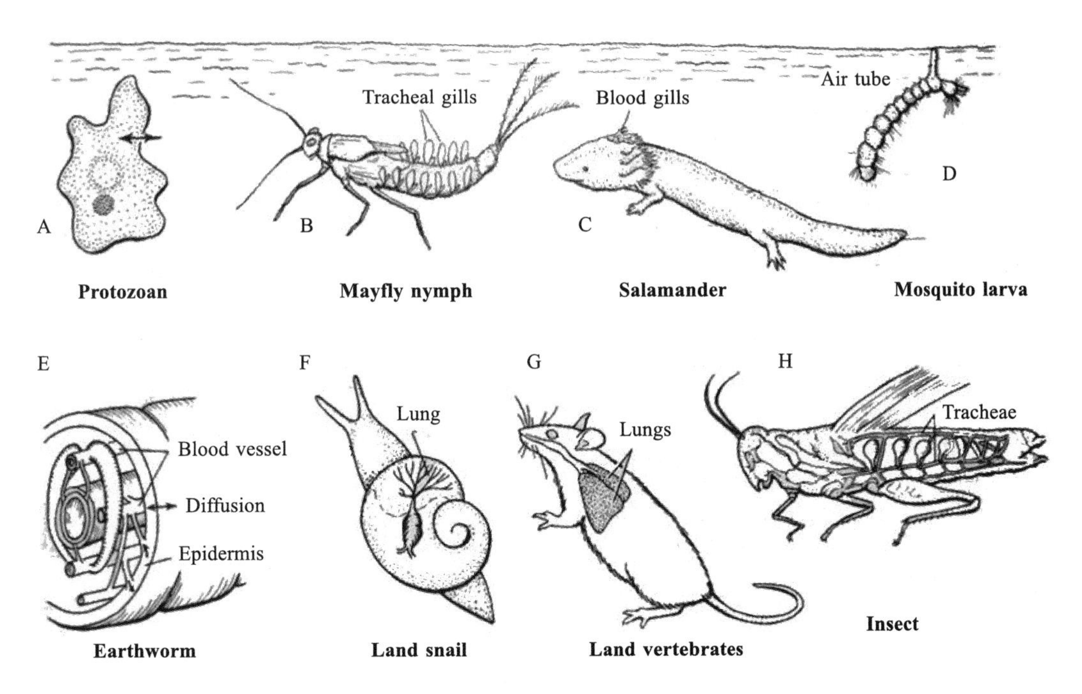

In water (above). A. Protozoan, diffusion through cell wall. B. Mayfly nymph (insect), tracheal gills. C. Salamander, blood gills. D. Mosquito larva, aquatic, with tube for breathing free air.
In air (below). E. Earthworm, diffusion through moist body wall to blood vessels. F. Land snail, moist lung inside body. G. Land vertebrate, pair of moist lungs inside body. H. Insect, system of air ducts (tracheae) throughout body.

Fig. 18-15 Types of respiratory mechanisms in animals (Storer et al., 1972)

The term "respiration" is normally associated with free oxygen. But some intestinal parasites and muck-inhabiting invertebrates live where there is little or no oxygen in their environment. These anaerobic animals may obtain energy in the absence of free oxygen through glycolysis.

The simplest-appearing method excretion is to pass wastes through the cell membrane into surrounding water, as occurs in many protozoans. Amoeba, *Paramecium*, and various other freshwater protozoans have one or more contractile vacuoles that accumulate excess water from within the cytoplasm and periodically discharge to the exterior so as to maintain the normal fluid balance within the cell body. Ammonia is the chief excretory product. The means of disposal of excretions by protozoans is obscure. Excretions of sponges and coelenterates diffuse from body cells into the epidermis and thence into the water. Among insects and a few other arthropods, the principal excretory organs are slender Malpighian tubules attached to the anterior end of the hindgut and closed at their free ends; these tubules collect wastes from the body fluids and discharge them into the hindgut. Both urates and carbon dioxide are received in solution from the blood water and other materials are resorbed in the lower parts of the tubules (Fig. 18-16). The final excretions, including uric acid crystals, carbonates, oxalates, and sometimes urea or ammonia, pass out with the feces. The fat body of insects is also a depository for organic wastes and is the chief excretory mechanism in springtails, which lack Malpighian tubules. The exoskeleton renders excretory service in some invertebrates, including insects,

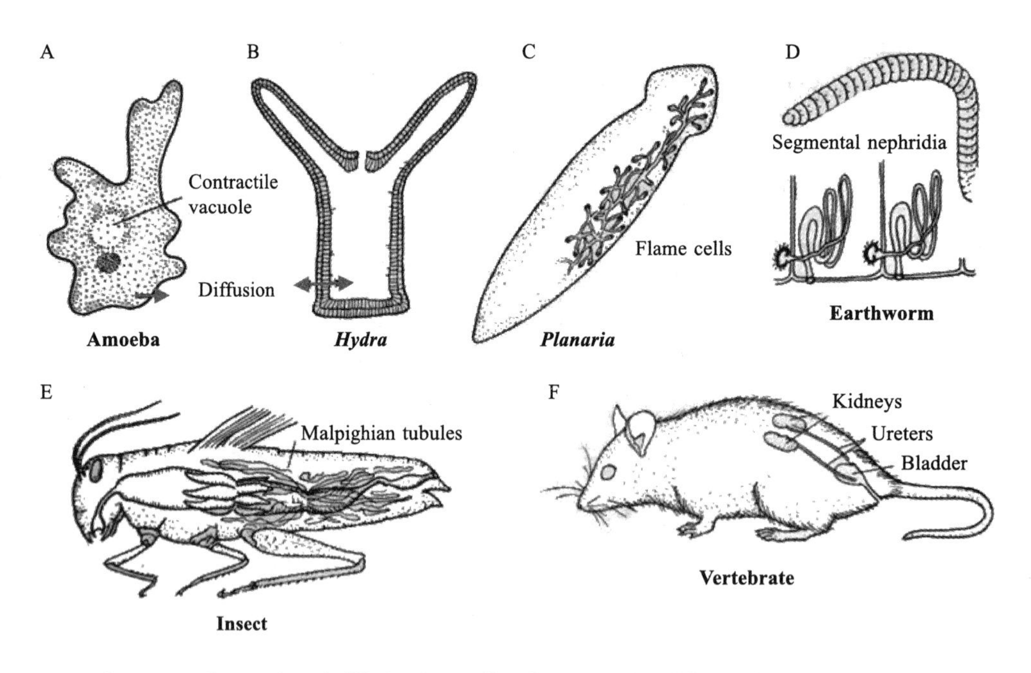

A. Amoeba, contractile vacuole and diffusion from cell surface. B. *Hydra*, diffusion from cells. C. *Planaria*, many flame cells connecting to ducts ending in common surface, excretory pore(s). D. Earthworm, two nephridia in each body somite, emptying separately through body wall. E. Grasshopper, series of fine Malpighian tubules connected to end of midgut. F. Vertebrate, two kidneys with ducts to a single bladder discharging to exterior.

Fig. 18-16 Types of excretory mechanisms in animals

(Storer et al., 1972)

since nitrogenous materials deposited in it are eliminated when the animal molts. The white pigment in wings of cabbage butterflies, formed from uric acid, is clearly an excretory product.

The commonest excretory organs in many animals are tubular structures, the nephridia and coelomoducts. Primitively these were arranged one pair to a body somite, but they have become variously modified in the course of evolution. Nephridia are ectodermal in origin. Flatworms and ribbon worms have many flame cells scattered among the body cells from which wastes are drawn to pass out in a branched system of ducts. In the earthworm, each somite contains a pair of nephridia. The inner end of each has a ciliated funnel, or nephrostome, draining from the coelom, and around the long tubule are blood vessels whence wastes are also drawn; the tubule ends externally as a small ventral nephridiopore. Mollusks and rotifers have one or two pairs of nephridium-like organs that drain from the body or blood; nephridia also occur in the chordate amphioxus.

In some annelids, mollusks, and arthropods and in the chordates, the principal excretory organs are coelomoducts, mesodermal in origin, probably derived from genital ducts, but now variously modified to remove wastes from the body cavity. Crustaceans have two pairs, the "antennal (green)" and "maxillary" glands; each has an end sac with a duct opening at the base of an appendage. Only rarely are both developed in the same stage of a single species. Spiders have coxal glands in the cephalothorax derived from coelomoducts.

18–9 Nervous Systems

All cells are excitable or irritable. Because of this, every organism is sensitive to changes or stimuli from both its external and its internal environments; to these it responds or reacts in various ways. Every type of organic response, from the simplest action of an amoeba to the most complex bodily function or mental process in man, results from this fundamental characteristic of excitability. To perceive stimuli, to transmit these to various body parts, and to effect responses, most animals have a nervous system. This system (together with endocrine glands) serves also to coordinate and integrate the functions of cells, tissues, and organ systems so that they act harmoniously as a unit (Fig. 18-17, Fig. 18-18).

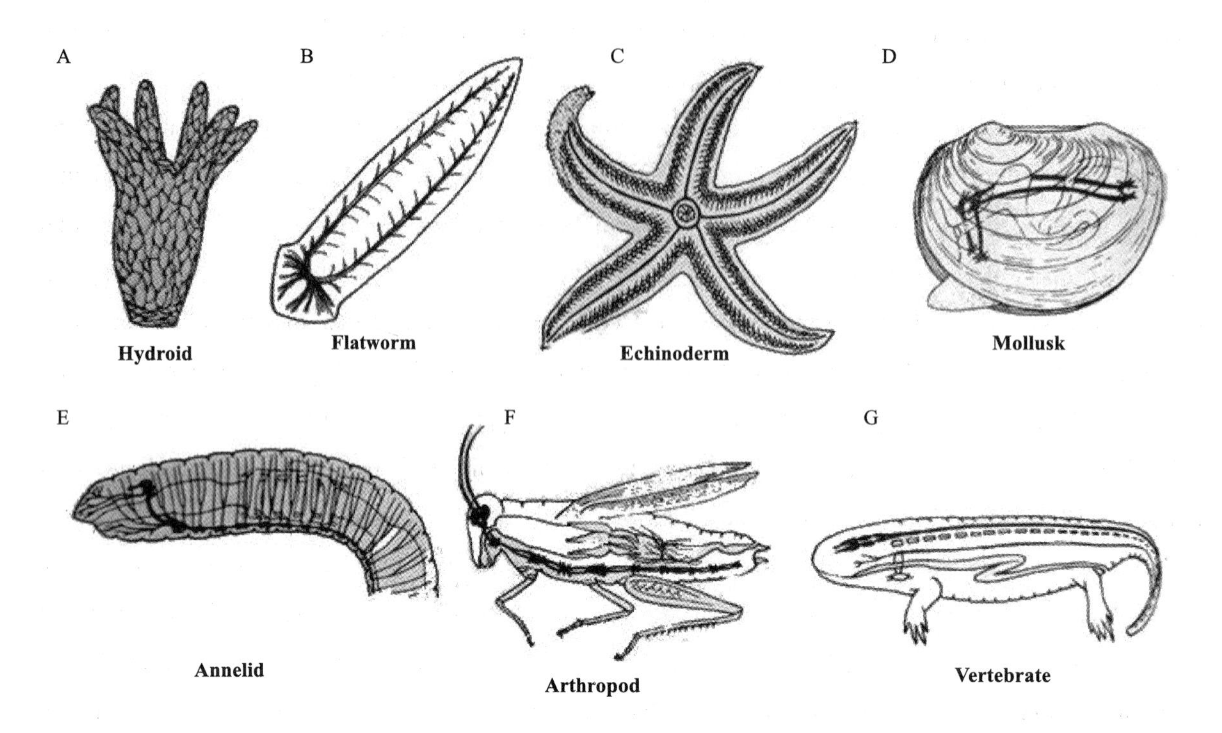

A. Hydroid, nerve net throughout body. B. Flatworm,ganglia in "head" region, two nerve cords. C. Echinoderm, nerve ring around mouth, median nerve in each arm. D. Bivalve mollusk, three pairs of ganglia and connectives. E. Annelid, a "brain" of ganglia in anterior end, double-solid ventral nerve cord, segmental ganglia and nerves. F. Arthropod, similar to earthworm. G. Vertebrate, brain in head, single hollow dorsal nerve cord with paired segmental nerves.

Fig. 18-17 Nervous systems (heavy black) in animals

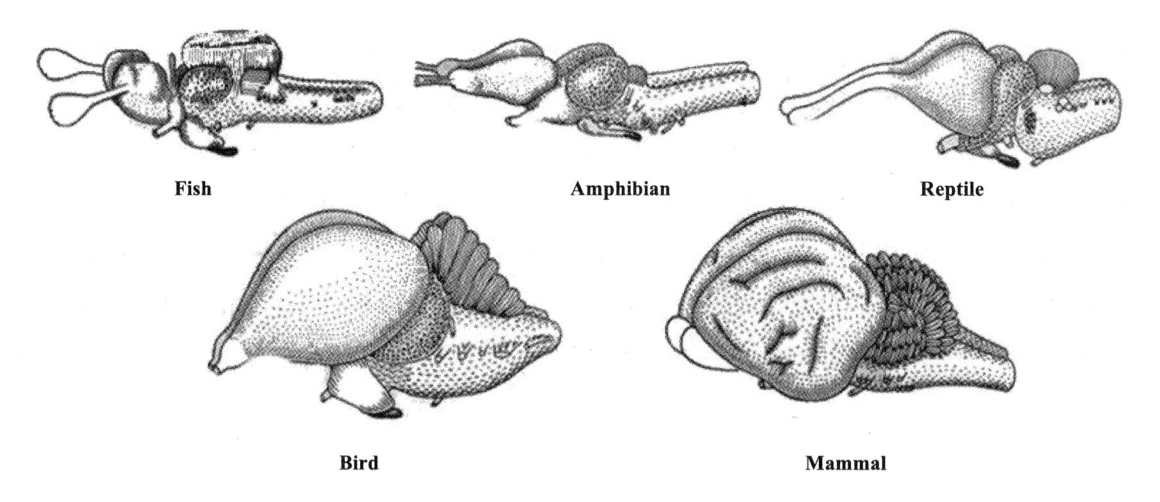

Olfactory lobes, clear; cerebrum, lightly stippled; optic tracts and lobes, coarsely stippled; base of midbrain, wavy lines; cerebellum, vertical lines; medulla oblongata, horizontal dashes; pituitary body, black. Stubs of cranial nerves are outlined.

Fig. 18-18 The brains of representative vertebrates showing progressive increase, especially in the cerebral hemispheres and cerebellum

18-10 学习要点

● 多细胞动物起源于单细胞动物的证据（古生物学、形态学、胚胎学）；胚胎学证据——生物发生律。

● 胚胎发育的重要阶段；关于多细胞动物起源的学说。

● 动物界进化历程。

● 动物系统的演化特点（皮肤、骨骼、肌肉、呼吸、排泄、消化、循环、神经、生殖）。

18-11 巩固测验

【名词】

中胚层、生物发生律、原生动物、后生动物、端细胞法、体腔囊法

【论述】

1. 归纳总结无脊椎动物体制的演化特征。

答题要点：对称性、体壁结构、体腔类型。

2. 归纳总结无脊椎动物呼吸相关结构的演化特征。

答题要点：体表渗透呼吸—水生物种的各种鳃—陆生物种的各种肺及器官。

3. 归纳总结无脊椎动物消化系统的进化特点。

答题要点：有口无肛门—有口有肛门—各门类的特殊性。

4. 归纳总结无脊椎动物神经系统的类型及演化。

答题要点：网状—梯形—筒状—索状—神经节。

5. 归纳总结无脊椎动物排泄系统的演化特点，列举排泄相关的细胞器或气管等。

答题要点：伸缩胞—原肾（管型、腺型）—后肾（马氏管等）。

6. 归纳总结无脊椎动物生殖与发育的多样性。

答题要点：无性生殖（出芽、包囊等）—有性生殖；世代交替—孤雌生殖等。发育主要指出现的幼虫及生活史不同阶段。

7. 归纳总结脊椎动物骨骼系统的演化特点。

答题要点：脊椎动物中轴骨的演化、特殊的骨骼（颌骨、颞孔、颈椎、荐椎等）。

8. 脊椎动物出现的具有里程碑意义的五个演化特征及其意义。

答题要点：上下颌、五趾型附肢、羊膜卵、恒温、胎生哺乳。

9. 绘制动物界系统进化树。

答题要点：由低等到高等，由原口动物到后口动物。

10. 陆栖脊椎动物呼吸系统演变的趋势。

答题要点：鳃—口咽式呼吸—胸腹式呼吸（气囊辅助双重呼吸）等。

11. 比较脊椎动物的心脏及动脉弓的演化。

答题要点：脊椎动物心房、心室、动脉弓的演变。

试题库

【试题一】

一、单项选择题（每小题2分，共10分）

1. 两栖类的呼吸方式是（　　）。

A. 口咽式　　B. 胸式

C. 腹式　　D. 胸腹式

2. 腔肠动物、环节动物、海绵动物以及扁形动物的幼虫分别是（　　）。

A. 浮浪幼虫、两囊幼虫、牟勒氏幼虫、担轮幼虫

B. 浮浪幼虫、牟勒氏幼虫、两囊幼虫、担轮幼虫

C. 浮浪幼虫、担轮幼虫、两囊幼虫、牟勒氏幼虫

D. 两囊幼虫、担轮幼虫、浮浪幼虫、牟勒氏幼虫

3. 水螅纲、珊瑚纲、钵水母纲的生殖腺分别来源于（　　）。

A. 外胚层、外胚层、内胚层　　B. 外胚层、内胚层、内胚层

C. 内胚层、外胚层、外胚层　　D. 内胚层、内胚层、外胚层

4. 下列具有一个枕髁的动物是（　　）。

A. 两栖类与爬行类　　B. 爬行类与鸟类

C. 鸟类与哺乳类　　D. 两栖类与哺乳类

5. 下列是裂体腔动物的是（　　）。

A. 蛔虫　　B. 环毛蚓

C. 涡虫　　D. 海星

二、填空题（每空1分，共10分）

以下分别是环毛蚓横切面图和蛙心脏简图，请对图中的结构进行标注。

①（　　）；②（　　）；③（　　）；④（　　）；⑤（　　）；⑥（　　）；⑦（　　）；⑧（　　）；⑨（　　）；⑩（　　）。

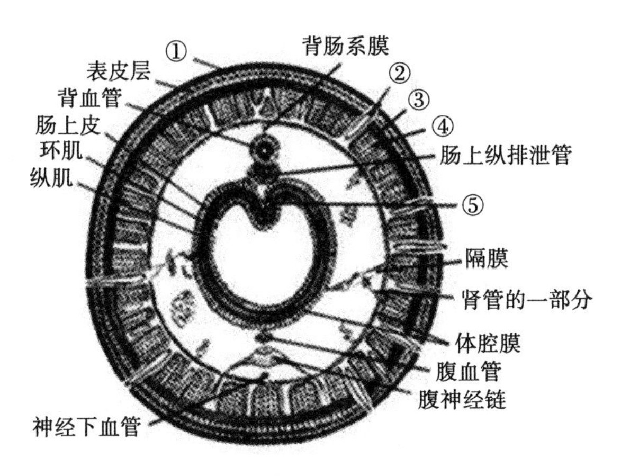

环毛蚓横切面

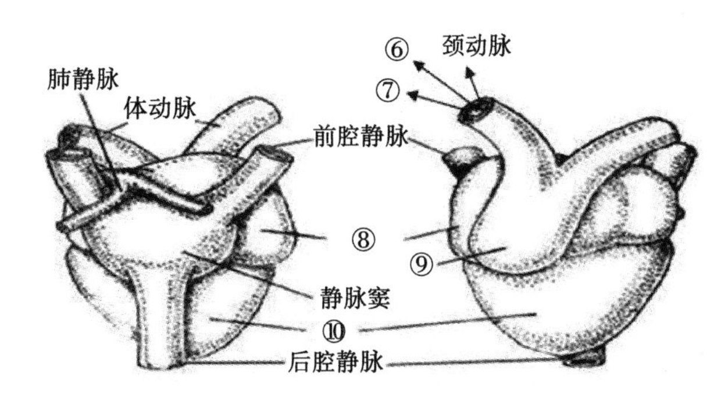

蛙的心脏背面观（左）和腹面观（右）

三、举例说明以下概念（每小题3分，共24分）

1. 水管系统　2. 水沟系　3. 马氏管　4. 外套膜
5. 海螵蛸　6. 开管式循环　7. 颞孔　8. 双重呼吸

四、简答题（每小题6分，共36分）

1. 软体动物门种类较多，请列出其中的4个纲并对每个纲举例1种动物。
2. 简述华支睾吸虫的生活史。
3. 比较真体腔和假体腔。
4. 举例说明昆虫口器的类型。
5. 简述脊椎动物呼吸系统演化特点。
6. 何为羊膜卵？羊膜卵的出现有何意义？

五、论述题（每小题10分，共20分）

1. 以环毛蚓和海鞘为例，说明脊索动物与原口动物主要的差异特征。
2. 绘制动物界系统进化树。

【试题二】

一、名词解释（每小题4分，共40分）

1. 赤潮　2. 群体学说　3. 两辐射对称　4. 黄色细胞　5. 茎化腕
6. 内柱　7. 鳃囊　8. 肝门静脉　9. 犁鼻器　10. 古北界

二、简答题（每小题10分，共40分）

1. 以虾为例，简述附肢的特化与其生活的适应性。

2. 简述动物（薮枝螅、轮虫、枝角类三类任选一类）的生活史。

3. 燕子等多种鸟类能在疾飞中捕食，这与其具有良好的视觉调节有关。简述鸟眼适应快速调节的结构及调节方式。

4. 简述鱼鳔的功能以及鳔内气体分泌和排放机制。

三、论述题（每小题15分，共30分）

1. 详述无脊椎动物门动物感受细胞器或器官的特点及其与生活习性的关系。

2. 以鱼、蛙和哺乳动物为例，谈谈水生动物从水生演化到陆生，血液循环系统发生的适应性变化。

四、实验题（每小题10分，共40分）

1. 简述活体蚯蚓解剖方法或过程，并绘制蚯蚓的横切面图。

2. 简述蟾蜍（蛙）解剖的方法或过程。如何从外部形态及性腺特征确定其性别？

3. 写出以下物种所属的门、纲，以及物种主要鉴别特征。

①草履虫 *Paramecium caudatum*；②猪带绦虫 *Taenia solium*；③圆田螺 *Cipangopaludina*；④园蛛 *Araneus diadematus*；⑤海盘车 *Asterias*。

4. 请写出下列动物的分类地位（目、科）及物种主要鉴别特征。

①鲢鱼 *Hypophthalmichthys molitrix*；②大黄鱼 *Pseudosciaena crocea*；③黑斑蛙 *Rana nigromaculata*；④喜鹊 *Pica pica*；⑤虎 *Panthera tigris*。

【试题三】

一、单项选择题（每小题1分，共10分）

1. 性成熟的环毛蚓在身体的14~16体节间无节间沟、状如指环的结构称为（　　）。

A. 体环　　B. 环带　　C. 卵茧　　D. 蚓茧

2. 寄生在人的淋巴管内的线虫是（　　）。

A. 丝虫　　B. 蛲虫　　C. 十二指肠钩虫　　D. 棘吻虫

3. 身体具有发光器，且胸肢全为双肢型的甲壳动物是（　　）。

A. 磷虾　　B. 糠虾　　C. 钩虾　　D. 沼虾

4. 柱头虫和脊索动物有较近的亲缘关系，因为它们的肌肉中都含有（　　）。

A. 磷肌酸　　B. 精氨酸　　C. 肌酸　　D. 谷氨酸

5. 消化道后端具有排泄腔和呼吸树的棘皮动物是（　　）。

A. 海参　　B. 海胆　　C. 海盘车　　D. 海羊齿

6. 水螅刺细胞中具有毒液的刺丝囊是（　　）。

A. 小型黏性刺丝囊　　B. 大型黏性刺丝囊

C. 缠卷刺丝囊　　D. 穿刺刺丝囊

7. 后口动物以体腔囊法形成中胚层和体腔，这种方式又称（　　）。

A. 肠体腔法　　B. 腔肠法　　C. 裂体腔法　　D. 端细胞法

8. 具有横纹肌的无脊椎动物是（　　）。

A. 扁虫　　B. 环节动物　　C. 软体动物　　D. 节肢动物

9. 纽形动物的幼虫称（　　）。

A. 纤毛幼虫　　B. 浮浪幼虫　　C. 牟勒氏幼虫　　D. 帽状幼虫

10. 人被蛔虫感染的原因是不小心吞食了它的（　　）。

A. 虫卵　　B. 囊蚴　　C. 胞蚴　　D. 尾蚴

二、填空题（每空1分，共20分）

1. 薮枝螅生活于浅海，为树状的水螅型群体，螅茎上分出两种个体：（　　　　）和（　　　　）。两者由（　　　　）中的（　　　　）相互连接。

2. 原生动物的营养方式主要有（　　　　）、（　　　　）和（　　　　）等三种方式。

3. 高等动物的组织分为（　　　　）组织、（　　　　）组织、（　　　　）组织和（　　　　）组织。

4. 人蛔虫体壁最内层是纵列的肌细胞，其结构可分为（　　　　）和（　　　　）两部分。

5. 环毛蚓雌雄同体，同律分节。其中，第14~16体节愈合，状如指环，称为（　　　　），其相当于第14体节腹面有一（　　　　）；第18体节腹面两侧具1对（　　　　）；另外，在第6和7、7和8和8和9节间沟腹面两侧有（　　　　）3对。

6. 华支睾吸虫生活史中历经（　　　　）个幼虫阶段，其中（　　　　）和（　　　　）能够进行幼体生殖。

三、名词解释（每小题4分，共40分）

1. 担轮幼虫　2. 脊索　3. 次生腭　4. 胸廓　5. 胞饮

6. 触腕　7. 鳃囊　8. 晚成雏　9. 动脉圆锥　10. 晶杆

四、简答题（每小题8分，共80分）

1. 试从发生顺序、所在位置和结构特征比较前肾、中肾和后肾的不同。

2. 什么是幼体生殖？以扁形动物为例说明寄生动物对寄生生活方式的适应。

3. 如何分辨出对虾的雌雄？对虾的附肢共有多少对？请写出它们的名称并简述其功能。

4. 举例说明同律分节和异律分节。

5. 什么是羊膜动物？说明羊膜卵在脊椎动物系统发生中的意义。

6. 什么是颌口类？颌口类主要包括哪些纲？请写出纲名及其主要特征。

7. 什么是“双名法”？写出动物分类7个阶元的中文、英文名称。

8. 根据海蜇的生活史说明世代交替现象。

9. 轮虫是重要的饵料动物。如对其开展人工培养，应解决何种生物学问题？说明原因。

【试题四】

一、名词解释（每小题5分，共50分）

1. 刺细胞　2. 胞饮（作用）　3. 世代交替　4. 两侧对称　5. 链式神经系统
6. 后肾　7. 皮肤肌肉囊　8. 孤雌生殖　9. 次生腭　10. 反刍

二、问答题（每小题10分，共100分）

1. 为什么说扁形动物在动物演化中具有重要地位？
2. 为什么鸟类能在空中飞行并在快速飞行中捕捉食物？
3. 何谓羊膜动物？羊膜卵的结构在动物演化中有什么意义？
4. 什么是颌口类？颌口类主要包括哪些纲？请写出纲名及其主要特征。
5. 什么是不完全变态？举例说明昆虫稚虫与若虫的特点。
6. 什么是后生动物？为什么说腔肠动物才是真正的两胚层动物？
7. 什么是原肾型排泄系统？其与后肾管在发生和结构上有什么不同？
8. 根据华支睾吸虫的生活史说明幼体生殖及其适应意义。
9. 论述海水和淡水水鱼类肾脏在调节体内渗透压方面所起的作用。
10. 说明蛔虫、蚯蚓、河蚌、虾、海星的体腔属于何种体腔？其血液循环的方式如何？

【试题五】

一、名词解释（每小题3分，共60分）

1. 奇静脉　2. 适应辐射　3. 胼胝体　4. 腔上囊　5. 侧线
6. 内柱　7. 附睾　8. 胸导管　9. 输精尿管　10. 开放式骨盆
11. 细胞器官　12. 辐射对称　13. 前胸腺　14. 书鳃　15. 血体腔
16. 螺旋卵裂　17. 不完全变态　18. 东洋区　19. 水管系统　20. 巨神经纤维

二、问答题（共90分）

1. 绘出家兔循环系统模式图，并注明各部位名称。（本小题10分）

2. 总结鸟类形态结构的主要特征。（本小题10分）

3. 总结脊椎动物亚门动物从水生过渡到陆生过程中，骨骼系统的演化趋势。（本小题10分）

4. 简述哺乳动物肾的结构及尿液形成的过程。（本小题10分）

5. 列举出科研中常用的给家兔和小鼠采血的方法。（本小题5分）

6. 比较日本血吸虫和蝗虫对各自生活方式的适应特点。（本小题15分）

7. 简述无脊椎动物呼吸器官（系统）的类型、结构和机能的演变。（本小题15分）

8. 举出5种中国珍稀脊椎动物（鱼类、两栖类、爬行类、鸟类和哺乳类各1种），并说明各自的地理分布。（本小题5分）

9. 举出5种海洋无脊椎动物发育中自由生活的幼虫期，说明其所属的门类和主要结构特征。（本小题10分）

【试题六】

一、名词解释（每小题2分，共20分）

1. 神经肌肉体系　2. 胞饮作用　3. 消化循环腔　4. 齿舌　5. 后口
6. 逆行变态　7. 鳃耙　8. 股孔或臀孔　9. 愈合荐骨　10. 群落

二、填空题（每空1分，共35分）

1. 一般认为动物体有4种基本组织，即（　　）、（　　）、（　　）和（　　）。

2. 关于多细胞动物起源有赫克尔的（　　）和梅契尼柯夫的（　　）两种学说。

3. 双翅目昆虫的（　　）退化成了（　　），这是无脊椎动物痕迹器官的典型例子。

4. 鱼类的洄游包括（　　）、（　　）和（　　）。

5. 羊膜卵的胚胎早期发育过程中，在胚胎外构成两个腔，即羊膜腔和胚外体腔。其中羊膜腔的壁称为（　　），而胚外体腔的壁称为（　　）。

6. 胚胎发育和胚后发育连接起来的全过程，称为（　　）。在胚后发育中，像两栖类那样有明显幼体期的，叫作（　　）；而像爬行类、鸟类等那样没有特殊的幼体期的，叫作（　　）。

7. 指出下图所示的结构名称：①（　　）；②（　　）；③（　　）；④（　　）；⑤（　　）；⑥（　　）；⑦（　　）。

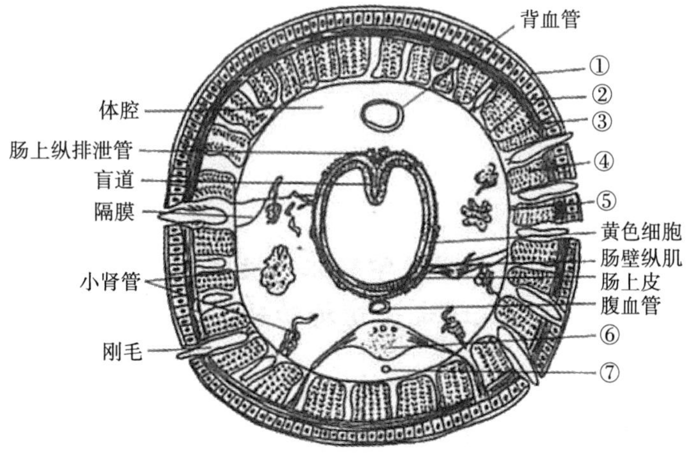

环毛蚓体中部横切图

8. 动物演变途径遵循一定的进化型式，一般进化型式可有（　　）、（　　）、（　　）、（　　）、（　　）和（　　）。

9. 脊索动物有一些性状也见于高等无脊椎动物，如（　　　　）、（　　　　）、（　　　　）、（　　　　）和（　　　　）等，这些共同点表明脊索动物由无脊椎动物演化而来。

10. 哺乳动物的子宫有多种类型，其原始类型为（　　　　）。

三、简答题（每小题10分，共50分）

1. 内耳、肾脏、真皮、毛、平滑肌各来源于哪个胚层?

2. 举例说明个体发育与系统发育的关系。

3. 概述鱼类受精和发育的几种类型，并举例说明。

4. 为什么说节肢动物由环节动物演变而来?

5. 简述物种的形成过程。

四、论述题（每小题15分，共45分）

1. 试从体制、分节、体腔、排泄和神经说明无脊椎动物的演化趋势。

2. 试述鱼类、两栖类、爬行类、鸟类和哺乳类的呼吸器官及呼吸方式。

3. 根据所学知识分析2007年6月洞庭湖地区鼠害暴发的原因及治理要点。

【试题七】

一、名词解释（每小题1.5分，共15分）

1. 扁形动物　2. 鳃　3. 鳞式　4. 尿囊　5. 生物发生律

6. 腔上囊　7. 内分泌腺　8. 卵胎生　9. 羊膜卵　10. 厌氧呼吸

二、填空题（每空1分，共35分）

1. 原生动物的主要纲有（　　　　）、（　　　　）、（　　　　）和（　　　　）。

2. 河蚌的排泄系统包括（　　　　）和（　　　　）。

3. 昆虫的小眼分为（　　　　）和（　　　　）两部分。

4. 鱼类的奇鳍一般包括（　　　　）、（　　　　）和（　　　　）3类。

5. 两栖动物包括（　　　　）目、（　　　　）目和（　　　　）目。

6. 鸟类和哺乳类的脊柱一般由（　　　　）、（　　　　）、（　　　　）、（　　　　）和（　　　　）5部分组成。

7. 哺乳动物的唾液腺包括（　　　　）腺、（　　　　）腺和（　　　　）腺。

8. 长江中生活的有（　　　　）豚和（　　　　）豚。

9. 颌弓与脑颅的连接一般有（　　　　）、（　　　　）和（　　　　）3种方式。

10. 淋巴系统包括（　　　　）、（　　　　）、（　　　　）和其他淋巴器官。

11. 脊椎动物的肾脏一般有（　　　　）、（　　　　）和（　　　　）3种类型。

12. 卵巢能分泌两种激素，一种是（　　　　），另一种是（　　　　）。

三、简答题（每小题10分，共50分）

1. 简述轮虫结构和机能的主要特点。

2. 从蚯蚓的结构特点简述其对土壤生活的适应性。

3. 简述河蚌在水产养殖和水环境治理中的作用。

4. 简述动物休眠的生物学意义。

5. 与鱼类相比，两栖类有哪些进步特征？

【试题八】

一、名词解释（每小题5分，共50分）

1. 弧胸型肩带　2. 皮肤肌肉囊　3. 颞窝　4. 咽颅　5. 洄游
6. 中生动物　7. 脊索　8. 蜕皮　9. 水管系统　10. 水沟系统

二、简答题（共100分）

1. 比较鸟类与哺乳类的骨骼系统，尤其是适应各自生活方式的不同特征。（20分）

2. 简述羊膜动物在繁殖方面的进步性特征。（10分）

3. 试述两栖类在其结构与功能方面适应陆生生活的特点及其不完善的方面。（20分）

4. 列举5种分别属于不同门、具有很强再生能力的动物，并写出这些动物所属的门和纲。（10分）

5. 简述节肢动物附肢的类型及其功能。（10分）

6. 寄生虫给寄主带来严重危害，举例论述寄生虫对寄主造成哪些危害。（20分）

7. 许多动物可以进行无性繁殖，但无性繁殖的方式也常不同。请举例说明动物有哪些无性繁殖的方式。（10分）

【试题九】

一、单项选择题（每小题2分，共20分）

1. 具有横纹肌的无脊椎动物是（　　）。

A. 扁虫　　B. 环节动物　　C. 软体动物　　D. 节肢动物

2. 下述消化道开始有肌肉组织的是（　　）。

A. 纽形动物　　B. 扁形动物　　C. 原腔动物　　D. 环节动物

3. 海绵动物体壁的内层由（　　）组成。

A. 领细胞　　B. 扁平细胞　　C. 内皮肌细胞　　D. 孔细胞

4. 下述属于后口动物的是（　　）。

A. 海参　　B. 东亚飞蝗　　C. 海葵　　D. 涡虫

5. 海鞘是近岸水域常见的污损生物，在系统分类上隶属于（　　）。

A. 半索动物　　B. 尾索动物　　C. 头索动物　　D. 腕足动物

6. 犁鼻器是（　　）。

A. 嗅觉器官　　B. 听觉器官　　C. 红外线感受器　　D. 压力感受器

7. 昆虫的排泄器官是（　　）。

A. 凯伯氏器　　B. 原肾管　　C. 后肾管　　D. 马氏管

8. 大鲵喜生活于山涧溪流的溶洞中，是变温动物，在系统发生上属于（　　）。

A. 无颌类和无羊膜类　　B. 有颌类和无羊膜类　　C. 有颌类和羊膜类　　D. 无颌类和羊膜类

9. 环节动物的消化道开始有肌肉组织，是由于其真体腔形成过程中具有（　　）。

A. 脏壁中胚层发生　　B. 体壁中胚层发生　　C. 体腔膜发生　　D. 肠系膜发生

10. 海葵的体制属于（　　）。

A. 不对称　　B. 两辐射对称　　C. 六辐射对称　　D. 八辐射对称

二、连线题（每小题1分，共10分）

1. 碟状体	A. 蚯蚓
2. 两囊幼虫	B. 海蜇
3. 牟勒氏幼虫	C. 田螺
4. 胞蚴	D. 对虾
5. 亚氏提灯	E. 蝗虫
6. 绿腺	F. 海绵
7. 生殖环带	G. 血吸虫
8. 马氏管	H. 海胆
9. 齿舌	I. 乌贼
10. 海螵蛸	J. 涡虫

三、名词解释（每小题2分，共20分）

1. 潘氏孔	2. 胞吞	3. 生物发生律	4. 晶杆	5. 包囊
6. 水管系统	7. 颞孔	8. 完全变态	9. 外套膜	10. 水沟系

四、简答题（每小题8分，共40分）

1. 说明鸟类适应飞翔生活的结构特征。
2. 腔肠动物的主要特征有哪些？分为哪几个纲？分别举例一种动物。
3. 脊索动物的主要特征有哪些？
4. 环节动物的主要特征有哪些？分为哪几个纲？分别举例一种动物。
5. 棘皮动物的主要特征有哪些？呼吸树有什么作用？

五、论述题（10分）

两栖动物适应陆地生活的特征有哪些？它们为什么离不开水环境？

【判断题】

1. 卵裂与细胞分裂不同，卵裂过程新的细胞未长大，又继续分裂，因此分裂成的细胞越来越小。

2. 囊胚期细胞还未开始分化，可以作为胚胎发育的干细胞。

3. 后口动物的棘皮动物、毛颚动物、半索动物及脊索动物均为体腔囊法形成胚层和体腔。

4. 生物发生律是一条客观规律，不仅适用于动物界，而且适用于整个生物界。

5. 海绵动物主要生活于海水中，全部营漂浮生活。

6. 海绵动物由于具有特殊的水沟系结构，故能很好地适应漂浮生活。

7. 海绵动物由于体表有许许多多的小孔，故又名多孔动物。

8. 海绵动物体表的一层细胞为领细胞，具保护作用。

9. 原始海绵体表穿插无数孔细胞，形成海绵的出水小孔。

10. 海绵动物体内为一层特有领细胞，能摄取食物进行细胞内消化。

11. 腔肠动物是一类比较高等的原生动物。

12. 水螅的体壁由两层细胞构成，在两层细胞之间为中胶层。

13. 由水螅内胚层分化而成的腺细胞，都可分泌黏液，有润滑作用。

14. 铁线虫属于线虫动物门。

15. 蛔虫雌雄异体。

16. 钩虫、丝虫和蛲虫都属于线虫动物。

17. 隐生是假体腔动物轮虫动物门所特有的习性。

18. 假体腔动物的排泄器官仍属于原肾管型。

19. 蛔虫为雌雄同体的动物。

20. 秀丽隐杆线虫是线虫动物门的模式动物，其揭示了动物细胞程序性死亡。

21. 蛔虫的雌虫有两个子宫和两个贮精囊。

22. 环节动物都须经过担轮幼虫期。

23. 沙蚕多数雌雄异体，蚯蚓多数雌雄同体。

24. 环节动物都具有生殖环带。

25. 沙蚕为同律分节。

26. 沙蚕和蚯蚓都具有真体腔。

27. 疣足是蚯蚓的运动器官。

28. 环节动物的排泄器官属于后肾管型。

29. 软体动物的数目超过节肢动物，是动物界第二大类群。

30. 软体动物一般为开管式循环，开管式循环的效率较低，一些快速游泳的种类例如乌贼，基本为闭管式循环。

31. 马氏管是陆生节肢动物的排泄器官。

32. 节肢动物具有几丁质的外骨骼。

33. 节肢动物身体属异律分节。

34. 节肢动物门昆虫纲是动物界种类最多的纲，其次是蛛形纲。

35. 倍足纲动物的足是多足纲动物的两倍。

36. 蝴蝶的触角类型是棒状。

37. 若虫是指不完全变态类昆虫幼虫。

38. 蝗虫为完全变态类昆虫。

39. 昆虫的口器基本类型是咀嚼式口器，其主要由上唇、上颚、下颚、下唇和舌五部分组成。

40. 棘皮动物是无脊椎动物中最高等的类群。

41. 区别海星的口面、反口面最明显的标志是在其反口面各腕中间均有一条由口伸向腕端部的步带沟。

42. 柄海鞘的血管无动脉和静脉之分，血液也无固定的单向流动方向。

43. 文昌鱼循环系统属于不完善的闭管式，无心脏，具有搏动能力的腹大动脉。

44. 脊索动物的背神经管起源于中胚层细胞，而脊索起源于内胚层细胞。

45. 头索动物一个显著的特征就是脊索终生存在，且达到头部。

46. 组成脊索或脊柱等内骨骼的细胞，能随同动物体发育而不断生长。

47. 鱼类和两栖类的皮肤都富有腺体。

48. 一般硬骨鱼类都具有喷水孔。

49. 绝大多数的鱼类鼻孔与口腔相通。

50. 鲤科是鱼类中种类最多的一个科。

51. 鱼类的间脑具有神经分泌的功能。

52. 鱼类的皮肤只起保护作用。

53. 硬骨鱼类的输精管由精巢外膜往后延续而成，与肾脏紧密相关。

54. 大多数鱼类是洄游性鱼类。

55. 蛙的口腔中有分泌黏液的颌间腺，分泌物具有湿润口腔和消化食物的作用。

56. 两栖类的生殖方式为体外受精。

57. 鱼类的眼无活动性眼睑，而大多数两栖动物的眼大而突出，具有活动性眼睑。

58. 两栖动物的皮肤裸露并富含腺体，鳞大多已退化，这是两栖动物区别于其他各纲脊椎动物的主要特征。

59. 两栖动物的膀胱除具有暂时贮存尿液的作用外，还具有重要的重吸收水分的机能。

60. 羊膜动物冬天都没有冬眠习性。

61. 大多数爬行动物所排尿液中的含氮废物为溶于水的尿素。

62. 蝰科、蝮亚科毒蛇位于眼与鼻孔之间的颊窝以及蟒科蛇类上颌前缘的唇窝，是一种极灵敏的热能感觉器官，属于红外线感觉器。

63. 爬行动物的眼都有活动性的上眼睑、下眼睑和瞬膜。

64. 蛇类适应穴居生活，中耳、鼓膜和耳咽管已退化，因此它不能感受空气传来的声波，只有感知从地面传来的声波。

65. 鸟类在小肠与大肠交界处着生有一对盲肠，以植物纤维为主食的鸟类（如鸡）盲肠特别发达。

66. 鸟类的主要消化腺是肝脏和胰脏，它们分别分泌胆汁和胰液注入空肠。

67. 气囊是保证鸟类在飞翔时供应足够氧气的装置。

68. 气体在鸟肺内沿一定方向流动，即背支气管—腹支气管—平行支气管。

69. 平胸总目和企鹅总目的鸟都不能飞，它们的胸骨突都不发达。

70. 哺乳动物处于应激状态时，交感神经兴奋；而副交感神经兴奋时，动物常处于放松休息的状态。

71. 不同于真兽亚纲动物，原兽亚纲和后兽亚纲动物的大脑皮层都不发达，也没有胼胝体。

72. 后兽亚纲的胚胎通过卵黄囊与母体子宫壁相接触，而真兽亚纲动物则是通过尿囊。

Vocabulary

◆ Chapter 1

amphibian /æm'fɪbiən/ *n.* 两栖动物；两栖类

archaea /ɑ:'ki:ə/ *n.* 古生菌（*sing.* archaeon /ɑ:'ki:ən/）

Arthropoda /ɑr'θrɒpədə/ *n.* 节肢动物门

as opposed to 与……截然相反；对照

Aves /'eivi:z/ *n.* 鸟纲，鸟类

bacteria /bæk'tɪəriə/ *n.* 细菌（*sing.* bacterium /bæk'tɪəriəm/）

behavioral ecology /bɪ'heɪvjərəl ɪ'kɒlədʒi/ *n.* 行为生态学

binomial nomenclature /baɪ'nəʊmiəl nəʊ'menkleɪtʃə(r)/ *n.* 双名法

botanist /'bɒtənɪst/ *n.* 植物学家

butterfly /'bʌtə(r)ˌflaɪ/ *n.* 蝴蝶；蝶泳

Canidae /'kænidi:/ *n.* 犬科；犬科动物

Carnivora /kɑ:'nivərə/ *n.* 食肉目；食肉动物

Chiroptera /kaiə'rɔptərə/ *n.* 翼手目

Chordata /kɔ:'deɪtə/ *n.* 脊索动物门；脊索动物

Chromista /'krəʊmɪstə/ *n.* 假菌界；藻物界；色藻界

cognition /kɒg'nɪʃ(ə)n/ *n.* 认识；认知；认识能力

coleopterology /ˌkɒlɪ'ɒptərɒlədʒi/ *n.* 鞘翅学

commonplace /'kɒmənˌpleɪs/ *n.* 老生常谈；*adj.* 平凡的，普通的

comparative anatomy /kəm'pærətɪv ə'nætəmi/ *n.* 比较解剖学

competition /ˌkɒmpə'tɪʃ(ə)n/ *n.* 竞争；比赛，竞赛

dorsal /'dɔ:(r)s(ə)l/ *adj.*（鱼或动物）背部的；*n.* 背部，背侧

ecology /ɪ'kɒlədʒi/ *n.* 生态学；社会生态学

ecosystem /'i:kəʊˌsɪstəm/ *n.* 生态系统

encompass /ɪn'kʌmpəs/ *v.* 包含；包围，环绕；完成

entomology /ˌentə'mɒlədʒi/ *n.* 昆虫学

equilibrium /ˌi:kwɪ'lɪbriəm/ *n.* 均衡；平衡

ethology /i:'θɒlədʒɪ/ *n.* 动物行为学

Eukaryota /juˌkærɪ'ɒtə/ *n.* 真核域；真核生物

eukaryotic /juˌkærɪ'ɒtɪk/ *adj.* 真核的，真核生物的

fertile /'fɜ:(r)taɪl/ *adj.* 富饶的，肥沃的；能生育的

fossil /'fɒs(ə)l/ *n.* 化石；僵化的事物；顽固不化的人；*adj.* 守旧的

fungi /'fʌŋgi:;'fʌŋgaɪ;'fʌndʒaɪ/ *n.* 真菌；菌类；蘑菇（*sing.* fungus /'fʌŋgəs/）

genus /'dʒi:nəs/ *n.* 属（*pl.* genera /'dʒenərə/）

geographic /ˌdʒi:ə'græfɪk/ *adj.* 地理的；地理学的

herpetology /ˌhɜ:pɪ'tɒlədʒɪ/ *n.* 爬虫学

homeostasis /ˌhəʊmiə'steɪsɪs/ *n.* 内稳态

hormone /'hɔ:(r)məʊn/ *n.* 激素

ichthyology /ˌɪkθɪ'ɒlədʒɪ/ *n.* 鱼类学；鱼类研究

Insecta /ɪn'sektə/ *n.* 昆虫纲；昆虫类

interbreed /ˌɪntə(r)'bri:d/ *v.*（使）混种；（使）异种交配

invertebrate /ɪn'vɜ:(r)tɪbrət/ *adj.* 无脊椎的；*n.* 无脊椎动物

lepidopterology /lepɪˌdɒptə'rɒlədʒɪ/ *n.* 鳞翅学

Mammalia /mæ'mɛliə/ *n.* 哺乳纲；哺乳类；哺乳动物

mammalogy /mæ'mælədʒɪ/ *n.* 哺乳动物学

mate /meɪt/ *n.* 配偶；配对物

microscope /'maɪkrəˌskəʊp/ *n.* 显微镜

moth /mɒθ/ *n.* 飞蛾，蛾

myrmecology /ˌmɜ:mɪ'kɒlədʒɪ/ *n.* 蚁学

offspring /'ɒfˌsprɪŋ/ *n.* 后代，子孙；产物

omnivore /'ɒmnɪˌvɔ:(r)/ *n.* 杂食者

ornithology /ˌɔ:(r)nɪ'θɒlədʒi/ *n.* 鸟类学

paleontology /ˌpælɪɒn'tɒlədʒɪ/ *n.* 古生物学

phylum /'faɪləm/ *n.* 门（分类）；语系（*pl.* phyla /'fəɪlə/）

physiology /ˌfɪzi'ɒlədʒi/ *n.* 生理学；生理机能

Plantae /'plænˌti:/ *n.* 植物界

Porifera /pɒ'rɪfərə/ *n.* 海绵动物门；海绵动物

porpoise /'pɔ:(r)pəs/ *n.* 海豚；鼠海豚

primatology /ˌpraɪmə'tɒlədʒɪ/ *n.* 灵长类动物学

Protista /prəʊ'tɪstə/ *n.* 原生生物界；原生生物

protozoa /ˌprəʊtə'zəʊə/ *n.* 原生动物（*sing.* protozoan /ˌprəʊtə'zəʊən/）

realm /relm/ *n.* 领域，范围；王国

relatedness /rɪ'leɪtɪdnəs/ *n.* 关联性；关系；亲缘

Reptilia /rep'tɪlɪə/ *n.* 爬行纲；爬行动物

sexuality /ˌsekʃu'æləti/ *n.* 性别；性欲；性征

skunk /skʌŋk/ *n.* 臭鼬

spatial distribution /'speɪʃ(ə)l ˌdɪstrɪ'bjuːʃ(ə)n/ *n.* 空间分布

sponge /spʌndʒ/ *n.* 海绵；海绵动物

subcategory /'sʌb'kætɪgərɪ/ *n.* 子分类；子范畴

subspecies /'sʌbˌspiːʃiːz; -ʃɪz; -spiːs-/ *n.* 亚种

Swedish /'swiːdɪʃ/ *adj.* 瑞典的；瑞典语的；瑞典人的；*n.* 瑞典语；瑞典人

taxonomy /tæk'sɒnəmi/ *n.* 分类学；分类法

taxonomic hierarchy /ˌtæksə'nɒmɪk 'haɪəˌrɑː(r)ki/ *n.* 分类层次结构

territory /'terət(ə)ri/ *n.* 领土，领域；范围；地域；版图

toad /təʊd/ *n.* 蟾蜍；癞蛤蟆；讨厌的家伙

vertebrate /'vɜː(r)tɪbrət/ *n.* 脊椎动物；*adj.* 有脊椎的

zoography /zəʊ'ɒgrəfɪ/ *n.* 描述动物学

zoology /zu'ɒlədʒi/ *n.* 动物学

◆ Chapter 2

acellular /ˌeɪ'seljələ(r)/ *adj.* 非细胞的

algae /'ældʒiː/ *n.* 藻类；海藻

alimentary /ælɪ'mentərɪ/ *adj.* 滋养的；食物的

amoeboid /ə'mibɒɪd/ *adj.* 变形虫样的

anal pore /'eɪn(ə)l pɔː(r)/ *n.* 肛孔

anisogamous /'ænɪsɒgəməs/ *adj.* 异形配子的

anterior /æn'tɪəriə(r)/ *adj.* 前面的；先前的

appendage /ə'pendɪdʒ/ *n.* 附件；附肢

asexually /eɪ'sekʃəlɪ/ *adv.* 无性地；无性生殖地

autogamy /ɔː'təʊgəmɪ/ *n.* 自体受精；自花授粉

autotrophic /ˌɔːtə'trɒfɪk/ *adj.* 无机营养的；自制养料的

binary fission /'baɪnəri 'fɪʃ(ə)n/ *n.* 二分裂

budding /'bʌdɪŋ/ *n.* 发芽；*adj.* 萌芽的；发育期的

canal /kə'næl/ *n.* 运河；管；*v.* 疏导

cilia /'sɪlɪə/ *n.* 纤毛；睫毛

Ciliata /'sɪlɪətə/ *n.* 纤毛纲；纤毛虫；纤毛亚纲

Ciliophora /sili'əufɔrə/ *n.* 纤毛亚门

coccidiosis /kɒkˌsɪdɪ'əʊsɪs/ *n.* 球虫病

commensal /kə'mensl/ *n.* 共生体；*adj.* 共生的

conjugation /ˌkɒndʒu'geɪʃ(ə)n/ *n.* 结合，配合

contractile /kən'træktaɪl/ *adj.* 可收缩的

corpse /kɔ:(r)ps/ *n.* 尸体

cramp /kræmp/ *n.* 痉挛，绞痛

creeping /'kri:pɪŋ/ *adj.* 爬行的；迟缓的

cuticle /'kju:tɪk(ə)l/ *n.* 角质层；表皮；护膜

cyst /sɪst/ *n.* 囊肿；包囊；膀胱

cytogamy /saɪ'təʊgəmɪ/ *n.* 细胞融合

cytoplasm /'saɪtəʊˌplæz(ə)m/ *n.* 细胞质

diarrhea /daɪə'ri:ə/ *n.* 腹泻，痢疾

digenetic /daɪdʒɪ'netɪk/ *adj.* 世代交替的，复殖的

digestion /daɪ'dʒestʃ(ə)n / *n.* 消化；领悟

diploid /'dɪplɔɪd/ *n.* 二倍体；*adj.* 二倍体的；双重的

duplicate /'dju:plɪkət/ *n.* 复制品；*v.* 倍增

ectoplasm /'ektəʊplæzəm/ *n.* 外质，外胚层质

Eimeria /ai'mi:riə/ *n.* 艾美球虫属

elastic /ɪ'læstɪk/ *adj.* 有弹性的；灵活的

encysted /en'sɪstɪd/ *adj.* 被囊的；包绕的

endomixis /endə'mɪksɪs/ *n.* 内融合

endoplasm /'endəʊplæzəm/ *n.* 内质；内胞浆

endoplasmic reticulum /endəʊ'plæzmɪk rɪ'tɪkjʊləm / *n.* 内质网

engulf /ɪn'gʌlf/ *vt.*吞没；吞食，狼吞虎咽

entangle /ɪn'tæŋg(ə)l/ *v.* 使纠缠；卷入；使混乱

epithelial /ˌepɪ'θi:lɪəl/ *adj.* 上皮的；皮膜的

Euglena /ju:'gli:nə/ *n.* 眼虫属

explode /ɪk'spləʊd/ *v.* 爆炸，爆发；激增

eyespot /'aɪˌspɒt/ *n.* 眼点；眼状斑点

fertilization /ˌfɜ:(r)təlaɪ'zeɪʃ(ə)n/ *n.* 施肥；受精

Flagellata /'flædʒəleɪtə/ *n.* 鞭毛动物纲

flagellum /flə'dʒeləm/ *n.* 鞭毛；鞭子

flexible /'fleksəb(ə)l/ *adj.* 灵活的；柔韧的

fusion /'fju:ʒ(ə)n/ *n.* 融合；融合物

gametogenesis /gæmətə'dʒenəsɪs/ *n.* 配子形成，配子发育

gametogony /ˌgæmɪ'tɒdʒəʊnɪ/ *n.* 配子生殖

granule /'grænju:l/ *n.* 颗粒

gullet /'gʌlɪt/ *n.* 食道；海峡；咽喉

habitat /'hæbɪtæt/ *n.* 栖息地，产地

haemolymph/'hi:məʊlɪmf/ *n.* 血淋巴

haploid /'hæplɔɪd/ *n.* 单倍体

hemixis /'hemɪksɪs/ *n.* 半融合

heterotrophic /ˌhetərə'trɒfɪk/ *adj.* 异养的

holophytic /ˌhɒlə(ʊ)'fɪtɪk/ *adj.* 自养的

holozoic /ˌhɒlə'zəʊɪk/ *adj.* 全动物营养的

host /həʊst/ *n.* 寄主

hypopharynx /ˌhaɪpə'færɪŋks/ *n.* 下咽部，喉咽

inchworm /'ɪn(t)ʃwɜ:m/ *n.* 尺蠖

inject /ɪn'dʒekt/ *v.* 注射

inoculate /ɪ'nɑkjəˌleɪt/ *v.* 给……接种

intracellular /ˌɪntrə'seljʊlə/ *adj.* 细胞内的

lemma /'lemə/ *n.* 膜

liberate /'lɪbəreɪt/ *v.* 解放；释放

locomotion /ˌləʊkə'məʊʃ(ə)n/ *n.* 运动；移动

longitudinal /ˌlɒndʒɪ'tju:dɪn(ə)l/ *adj.* 长度的，纵向的；经线的

macronucleus /mækrəʊ'nju:klɪəs/ *n.* 滋养核，大核

malaria /mə'leəriə/ *n.* 疟疾；瘴气

Mastigophora /mæsti'gɒfərə/ *n.* 鞭毛纲

merozoite /merə'zəʊaɪt/ *n.* 裂殖子

metabolism /mə'tæbəˌlɪz(ə)m/ *n.* 新陈代谢

micronucleus /maɪkrəʊ'nju:klɪəs/ *n.* 微核；小核

microorganism /ˌmaɪkrəʊ'ɔ:(r)gənɪz(ə)m/ *n.* 微生物；微小动植物

mitochondria /ˌmaɪtəʊ'kɒndrɪə/ *n.* 线粒体

mitosis /ˌmaɪ'təʊsɪs/ *n.* 有丝分裂

monogenetic /mɒnɒdʒɪ'netɪk/ *adj.* 一元发生的；无性生殖的

mutually /'mju:tʃuəli/ *adv.* 互相地

nourishment /'nʌrɪʃmənt/ *n.* 食物；营养品

nucleus /'nju:kliəs/ *n.* 核，核心；原子核

nutrition /nju'trɪʃ(ə)n/ *n.* 营养；营养学；营养品

oocyst /'əʊəsɪst/ *n.* 卵囊

oral groove /'ɔ:rəl gru:v/ *n.* 口沟；围口部

organelle /ˌɔ:(r)gə'nel/ *n.* 细胞器

osmoregulation /ɒzməʊregjʊ'leɪʃ(ə)n/ *n.* 渗透调节

osmosis /ɒz'məʊsɪs/ *n.* 渗透；渗透作用

Paramecium /ˌpærə'mi:sɪəm/ *n.* 草履虫属

parasitic /ˌpærə'sɪtɪk/ *adj.* 寄生的

parental /pə'rent(ə)l/ *adj.* 父母亲的；亲本的

pellicle /'pelɪk(ə)l/ *n.* 薄膜，薄皮

phagocytosis /ˌfægəsaɪ'təʊsɪs/ *n.* 吞噬作用

phagotrophy /'fægətrəfi/ *n.* 吞噬

photosynthesis /ˌfəʊtəʊ'sɪnθəsɪs/ *n.* 光合作用

pinch /pɪntʃ/ *v.* 捏；夹紧；挤压；收缩

piroplasm /pɪrə'plæzəm/ *n.* 梨浆虫

plasmodium /plæz'məʊdɪəm/ *n.* 疟原虫；变形体；原形体；多核的原形质块

Plasmodium vivax /plæz'məʊdɪəm 'vaivæks/ *n.* 间日疟原虫

plural /'plʊərəl/ *adj.* 复数的；多样的

projection /prə'dʒekʃ(ə)n/ *n.* 突出

protist /'prəutɪst/ *n.* 原生生物

protoplasm /'prəʊtəplæz(ə)m/ *n.* 原生质；细胞质

pseudopod /'sju:də(ʊ)pɒd/ *n.* 伪足；假足

puddle /'pʌd(ə)l/ *n.* 水坑；泥潭

reproduction /ˌri:prə'dʌkʃ(ə)n/ *n.* 繁殖，生殖

reservoir /'rezə(r)ˌvwɑ:(r)/ *n.* 水库；蓄水池

respiration /ˌrespə'reɪʃ(ə)n/ *n.* 呼吸；呼吸作用

ribosome /'raɪbəsəʊm/ *n.* 核糖体；核蛋白体

salivary /sə'laɪvəri/ *adj.* 唾液的

saprophytic /sæprə'fitik/ *adj.* 腐生的

scrunching /skrʌntʃ/ *v.* 把……揉成一团

schizogony /skɪ'zɒgənɪ/ *n.* 分裂生殖

sickle-shaped /'sɪklʃeɪpt/ *adj.* 镰形的

slender /'slendə(r)/ *adj.* 细长的；苗条的

specimen /'spesɪmɪn/ *n.* 样品，样本；标本

sporogony /spɒ'rɒgənɪ/ *n.* 孢子生殖，孢子发生

sporozoa /spɒrəu'zəuə/ *n.* 孢子虫类

sporozoite /ˌspɒrə(ʊ)'zəʊaɪt/ *n.* 孢子体

syngamy /'sɪŋgəmɪ/ *n.* 两性生殖；配子配合

toxin /'tɒksɪn/ *n.* 毒素，毒质；毒素类

transverse /'trænzvɜ:(r)s/ *n.* 横断面；*adj.* 横向的
trichocyst /'trɪkəsɪst/ *n.* 刺细胞
trophozoite /trɒfə'zəuɪt/ *n.* 滋养了；营养体
unicellular /ˌju:nɪ'seljʊlə(r)/ *adj.* 单细胞的
vacuole /'vækjuəʊl/ *n.* 液泡；空泡
vesicle /'vesɪkl/ *n.* 囊泡；小水泡
wrap /ræp/ *vt.* 包；缠绕；隐藏；掩护
zygote /'zaɪgəʊt/ *n.* 受精卵；接合子

◆ Chapter 3

accomplish /ə'kʌmplɪʃ/ *v.* 完成；实现；达到
amphiblastula /æmfəblæst'jʊlə/ *n.* 两囊幼虫
ascon /'æskɒn/ *n.* 单沟型
asymmetrical /ˌeɪsɪ'metrɪk(ə)l/ *adj.* 非对称的
calcareous /kæl'keərɪəs/ *adj.* 钙质的，石灰质的
calcium /'kælsiəm/ *n.* 钙
carbonate /'kɑ:bəneɪt/ *n.* 碳酸盐
chamber /'tʃeɪmbə(r)/ *n.* 腔隙；腔；心室；内庭
choanocyte /kəʊ'ænəsaɪt/ *n.* 环细胞，领细胞
choanoderm /kəʊ'ænədəm/ *n.* 领细胞膜
choanoflagellate /'kəʊnɑ:flædʒəleɪt/ *n.* 领鞭虫类
collar /'kɒlə(r)/ *n.* 颈部
colony /'kɒləni/ *n.* 种群；动物栖息地
Desmospongia /ˌdemə'spɒndʒiə/ *n.* 普通海绵纲
embryological /ˌembrɪə'lɒdʒɪkl/ *adj.* 胚胎学的
excurrent /eks'kʌrənt/ *n.* 出水管
filtration /fɪl'treɪʃ(ə)n/ *n.* 过滤
gemmule /'dʒemju:l/ *n.* 小芽；芽球；无性芽
hermaphrodite /hɜ:(ə)'mæfrədaɪt/ *n.* 雌雄同体
Hexactinellida /ˌheksə'tinəlɪdə/ *n.* 六放海绵纲
invagination /ɪnˌvædʒɪ'neɪʃən/ *n.* 内陷
larva /'lɑ:(r)və/ *n.* 幼体，幼虫
leucon /'lju:kən/ *n.* 复沟型
maternal /mə'tɜ:(r)n(ə)l/ *adj.* 母性的；母系的
mesenchymal /mes'eŋkɪməl/ *adj.* 间叶细胞的

mesohyl /'mesəʊhɪl/ *n.* 中胶层

metamorphose /ˌmetə'mɔ:fəʊz/ *v.* 完全变态

microfilament /'maɪkrəʊˌfɪləmənt/ *n.* 细胞微丝

microvilli /ˌmaɪkrəʊ'vɪlaɪ/ *n.* 微绒毛（*sing.* microvillus /ˌmaɪkrəʊ'vɪlas/）

multicellular /ˌmʌltɪ'seljələ(r)/ *adj.* 多细胞的

neurosensory /ˌnjʊərəʊ'sensərɪ/ *adj.* 感觉神经的

osculum /'ɒskjʊləm/ *n.* 排水孔；吸盘

ostia /'ɒstɪə/ *n.* 口；门；孔（*sing.* ostium /'ɒstɪəm/）

ova /'əʊvə/ *n.* 卵细胞（*sing.* ovum /'əʊvəm/）

oval /'əʊv(ə)l/ *adj.* 卵形的

parenchymula /pæreŋ'kɪmjʊlə/ *n.* 中实幼虫

pinacocyte /ˌpɪnə'kəʊsaɪt/ *n.* 扁平细胞

poriferan /pə'rɪf(ə)rən/ *n.* 多孔动物；海绵动物

porocyte /pɔ:'rɒsaɪt/ *n.* 孔细胞

regeneration /rɪˌdʒenə'reɪʃn/ *n.* 再生；重建

Scypha /'saifə/ *n.* 樽海绵属

sperm /spɜ:m/ *n.* 精子；精液

spicule /'spɪkju:l/ *n.* 骨针，骨片

spongin /'spʌndʒɪn/ *n.* 海绵硬蛋白

spongocoel /'spɒŋgəʊsi:l/ *n.* 海绵腔

substratum /'sʌbstrɑ:təm/ *n.* 基质；底质

sycon /'saɪkɒn/ *n.* 双沟型

symbiotic /ˌsɪmbaɪ'ɒtɪk/ *adj.* 共生的；共栖的

symmetrical /sɪ'metrɪk(ə)l/ *adj.* 对称的

◆ Chapter 4

acraspedote /ə'kræspəˌdəʊt/ *adj.* 无缘膜的

anatomical /ˌænə'tɒmɪkl/ *adj.* 解剖的；解剖学的

anemone /ə'neməni/ *n.* 海葵；银莲花

Anthozoa /ˌænθə'zəuə/ *n.* 珊瑚虫类

anus /'eɪnəs/ *n.* 肛门

aperture /'æpətʃə(r)/ *n.* 孔，穴

barb /bɑ:b/ *n.* 羽支；倒刺

biradial /baɪ'reɪdɪəl/ *n.* 两侧辐射对称的

blastula /'blæstjʊlə/ *n.* 囊胚；囊胚泡

canopy /'kænəpi/ *n.* 天篷；华盖；遮篷

carnivore /kɑ:nɪvɔ:(r)/ *n.* 食肉动物；食虫植物

circlet /'sɜ:klət/ *n.* 小圈

clownfish /'klaʊnfɪʃ/ *n.* 小丑鱼；海葵鱼

cnidarian /naɪ'deərɪən/ *n.* 刺胞动物

cnidocil /knɪdəʊ'sɪl/ *n.* 刺针；刺细胞突起

cnidocyte /'knɪdəʊsaɪt/ *n.* 刺细胞

coalesce /ˌkəʊə'les/ *v.* 合并；结合；联合

Coelenterata /si:'lentəreitə/ *n.* 腔肠动物门

coelenteron /sɪ'lentərɒn/ *n.* 腔肠动物的腔肠体

colloquially /kə'ləʊkwiəli/ *adv.* 口语地

complexity /kəm'pleksəti/ *n.* 复杂性

configuration /kənˌfɪgə'reɪʃn/ *n.* 配置；结构

craspedote /'kræspəˌdəʊt/ *n.* 缘膜

cross-section /'krɒs sekʃn/ *n.* 横截面，横断面

cylindrical /sə'lɪndrɪkl/ *adj.* 圆柱形的；圆柱体的

dioecious /daɪ'i:ʃəs/ *adj.* 雌雄异体的

dioxide /daɪ'ɒksaɪd/ *n.* 二氧化物

diploblastic /ˌdɪpləʊ'blæstɪk/ *adj.* 双胚层的

distal /'dɪstl/ *adj.*末梢的；末端的

diverticula /daɪvə'tɪkjʊlə(r)/ *n.* 盲囊；支囊（*sing.* diverticulum /ˌdaɪvə'tɪkjʊləm/）

dome /dəʊm/ *n.* 圆屋顶；半球形物

ectoderm/'ektəʊˌdɜ:m/ *n.* 外胚层

egest /ɪ:'dʒest/ *vt.* 排泄；排出（汗、粪便等）

endoderm /'endə(ʊ)dɜ:m/ *n.* 内胚层

ensnare /ɪn'sneə(r)/ *v.* 诱捕

ephyra /'efərə/ *n.* 蝶状幼体

epidermis /ˌepɪ'dɜ:mɪs/ *n.* 上皮；表皮

excitatory /ɪk'saɪtət(ə)rɪ/ *adj.* 兴奋的；刺激性的

excretory /ɪks'kri:təri/ *n.* 排泄器官；*adj.* 排泄的；分泌的

expulsion /ɪk'spʌlʃn/ *n.* 呼出

flesh /fleʃ/ *n.* 肉；果肉；肉色

fragmentation /ˌfrægmen'teɪʃn/ *n.* 破碎；分裂

gastrodermis /'gæstrədɜ:mɪs/ *n.* 肠表皮

gastrovascular /ˌgæstrəʊ'væskjʊlə(r)/ *adj.* 消化与循环两用的

gastrozooid /ˌgæstrə'zəʊɒɪd/ *n.* 营养个体

gonozooid /'gɒnəˌzəʊɒɪd/ *n.* 生殖虫体

hermatypic /ˌhɜ:mə'tɪpɪk/ *adj.* 造礁的

hermit /'hɜ:mɪt/ *n.* 蜂鸟

hollow /'hɒləʊ/ *adj.* 中空的，空腹的；凹的

holoblastic /ˌhɒlə'blæstɪk/ *adj.* 完全分裂的

Hydrozoa /ˌhaɪdrə'zoə/ *n.* 水螅纲

hydrula /haɪ'drʊlə/ *n.* 螅状幼体；原芽体

immobilize /ɪ'məʊbəlaɪz/ *v.* 使固定

manubrium /mə'nju:brɪəm/ *n.* 柄状体

medusa /mɪ'dju:zə/ *n.* 水母

mesentery /'mesəntərɪ/ *n.* 肠系膜

mutualistic /'mju:tʃʊə'lɪstɪk/ *adj.* 共生的

necessitating /nə'sesɪteɪtɪŋ/ *v.* 需要；迫使

nematocyst /'nemətəˌsɪst/ *n.* 刺细胞；刺丝囊

ocelli /əʊ'selaɪ/ *n.* 单眼类昆虫或动物

Obelia /əʊ'bi:ljə/ *n.* 薮枝虫；薮枝螅

penetrate /'penətreɪt/ *v.* 渗透；穿透

peptide /'peptaɪd/ *n.* 肽

pharyngeal /fə'rɪndʒiəl/ *adj.* 咽的；*n.* 喉音

pharynx /'færɪŋks/ *n.* 咽

photosynthetic /fəʊtəʊsɪn'θetɪk/ *adj.* 光合作用的

Physalia /faɪ'seɪlɪə/ *n.* 僧帽水母属

planula /'plænjʊlə/ *n.* 浮浪幼虫

plexus /'pleksəs/ *n.*（血管、淋巴管、神经等的）丛

polymorphic /ˌpɒlɪ'mɔ:fɪk/ *adj.* 多态的；多形的

polyp /'pɒlɪp/ *n.* 珊瑚虫；水螅虫

polypoid /'pɒlɪpɔɪd/ *adj.* 水螅体的，水螅似的

rhopalia /rəʊ'peɪlɪə/ *n.* 感觉棍（*sing.* rhopalium /rəʊ'peɪlɪəm/）

rudiment /'ru:dɪm(ə)nt/ *n.* 雏形；退化器官

sac /sæk/ *n.* 囊，液囊

scyphistoma /saɪ'fɪstəmə/ *n.* 钵口幼体

Scyphozoa /ˌsaifə'zəuə/ *n.* 钵水母纲

sea pen /si: pen/ *n.* 海鳃

septa /'septə/ *n.* 隔膜；隔板（*sing.* septum /'septəm/）

sessile /'sesaɪl/ *adj.* 固着的

siphonoglyph /saɪ'fɒnəglɪf/ *n.* 口道沟

solitary /'sɒlətri/ *adj.* 单独的

stalk /stɔ:k/ *n.*（植物的）茎，秆

statocyst /'stætəʊsɪst/ *n.* 平衡囊；平衡胞

stinger /'stɪŋə/ *n.* 螯针；毒刺

stomodaeum /ˌstəʊmə'di:əm/ *n.* 口道；原口

subdue /səb'dju:/ *vt.* 征服；抑制；减轻

tentacle /'tentəkl/ *n.* 触手，触须，触角

tetramerous /te'træmərəs/ *adj.* 四部分的

trough /trɒf/ *n.* 海槽；水槽；波谷

Velella /ve'lelə/ *n.* 帆水母属

velum /'vi:ləm/ *n.* 缘膜

venomous /'venəməs/ *adj.* 有毒的；分泌毒液的

zooxanthellae /zəʊəʊzæn'θeli:/ *n.* 黄藻

◆ Chapter 5

acoelomate /eɪ'si:ləˌmeɪt/ *n.* 无体腔动物；*adj.* 无体腔的

adjacent/ə'dʒeɪsnt/ *adj.* 邻近的；毗连的

anus /'eɪnəs/ *n.* 肛门

bilateral /ˌbaɪ'lætərəl/ *adj.* 双边的；有两边的

bilharzia /bɪl'hɑ:tsiə/ *n.* 血吸虫；血吸虫病

bilobed /bai'ləʊbd/ *adj.* 二裂片的

burrow /'bʌrəʊ/ *n.*（动物的）洞穴，地洞；*v.* 掘地洞

cerebral /sə'ri:brəl/ *adj.* 大脑的，脑的

Cestoda /'sestədə/ *n.* 多节绦虫亚纲，绦虫类

Clonorchis sinensis /ˌkləu'nɔ:kis 'sininsis/ *n.* 华枝睾吸虫

cross-fertilization /ˌkrɒsˌfɜ:rtələ'zeɪʃn/ *n.* 异花受精

Digenea /dai'dʒeniə/ *n.* 复殖亚纲；复殖类

Diphyllobothrium latum /daiˌfi:lə'bəuθriəm 'leɪtəm/ *n.* 阔节裂头绦虫

dorsoventrally /ˌdɔ:sə(ʊ)'ventr(ə)li/ *adv.* 背腹地

Dugesia japonica /du'dʒɪsiəˌdʒə'pɒnɪkə/ *n.* 日本三角涡虫

dwarf /dwɔ:f/ *adj.* 矮小的；矮生的

encapsulating /en'kæpsəˌleɪtɪŋ/ *n.* 封装

encyst /ɪn'sɪst/ *n.* 被囊；包囊

feces /'fi:si:z/ *n.* 排泄物，粪便；渣滓

fission /'fɪʃn/ *n.* 裂变；分裂；分裂生殖法

flame /fleɪm/ *n.* 火焰

flatworm /'flætwɜ:m/ *n.* 扁形虫

fluke /flu:k/ *n.* 吸虫

follicle /'fɒlɪkl/ *n.* 卵泡；滤泡；小囊

ganglia /'gæŋglɪə/ *n.* 神经节；神经中枢

germarium /dʒɜ:'meərɪəm/ *n.* 生殖腺，生殖室；卵巢

glide /glaɪd/ *v.* 滑翔；滑行

gravid /'grævɪd/ *adj.* 妊娠的；怀孕的

hatch /hætʃ/ *n.* 孵化；*v.* 孵

hook /hʊk/ *n.* 挂钩；钩状器官（或突起），倒刺

Hymenolepis nana /ˌhaimə'nɒləpis 'nɑ:nə/ *n.* 短膜壳绦虫

infest /ɪn'fest/ *vt.* 侵染；感染；寄生于

intermediate /ˌɪntə'mi:diət/ *adj.* 中间的，过渡的；*n.*（化合物）中间体

isthmus /'ɪsməs/ *n.* 地峡；峡部

knob /nɒb/ *n.* 球形突出物

lobe /ləʊb/ *n.* 耳垂；脑叶

mucous /'mju:kəs/ *adj.* 黏液的；分泌黏液的

ootype /'əʊətaɪp/ *n.* 卵模（腔）

opaque /əʊ'peɪk/ *adj.* 不透明的；*n.* 不透明物

oviduct /'əʊvɪdʌkt/ *n.* 输卵管

parasite /'pærəsaɪt/ *n.* 寄生虫；寄生生物

planarian /plə'neərɪən/ *n.* 真涡虫；三肠涡虫

Platyhelminthes /ˌplætɪhel'mɪnθiz/ *n.* 扁形动物门；扁形动物

proglottid /prəʊ'glɒtɪd/ *n.* 节片

proliferate /prə'lɪfəreɪt/ *v.* 增殖；扩散；激增

propel /prə'pel/ *v.* 推进；驱使；激励；驱策

protonephridia /ˌprəʊtəʊnɪ'frɪdɪə/ *n.* 原肾管（*sing.* protonephridium /ˌprəʊtəʊnɪ'frɪdɪəm/）

rostellum /rɒ'steləm/ *n.* 顶突；小喙

scavenger /'skævɪndʒə(r)/ *n.* 食腐动物；清道夫

Schistosomatidae /'ʃɪstə(ʊ)ˌsəʊ'mɪtidi/ *n.* 裂体科

schistosomiasis /ˌʃɪstə(ʊ)səʊ'maɪəsɪs/ *n.* 血吸虫病

scolex /'skəʊleks/ *n.* 头节

securing /sɪ'kjʊriŋ/ *adj.* 固定住的

segmented /seg'mentɪd/ *adj.* 分段的；*v.* 分割

self-fertilization /ˌself fɜ:təlaɪ'zeɪʃn/ *n.* 自花受精

shed /ʃed/ *v.* 摆脱；使落下；流出；脱落；（动物）蜕（皮），脱（毛）

strobila /strə'baɪlə/ *n.* 横裂体

sucker /'sʌkər/ *n.* 吸盘

Taenia solium /'ti:nɪə ˌsəʊlɪəm/ *n.* 猪肉绦虫

tapeworm /'teɪpwɜ:m/ *n.* 绦虫

Trematoda /ˌtremə'təudə/ *n.* 吸虫纲

triploblastic /ˌtrɪplə(ʊ)'blæstɪk/ *adj.* 三胚层的

Turbellaria /ˌtɜ:bə'leərɪə/ *n.* 涡虫纲

uterus /'ju:tərəs/ *n.* 子宫

vagina /və'dʒaɪnə/ *n.* 阴道；生殖腔

ventral /'ventrəl/ *adj.* 腹侧的；腹部的

vitelline /vɪ'telɪn/ *adj.* 卵黄的

◆ Chapter 6

acanthocephalan /əˌkænθə'sefələn/ *n.* 棘头虫

arthropod /'ɑ:θrəpɒd/ *n.* 节肢动物

ascend /ə'send/ *v.* 攀登，上升

Aschelminthes /æs'tʃəlminθs/ *n.* 囊蠕虫门

blastocoel /'blæstəusi:l/ *n.* 囊胚腔；卵裂腔；分裂腔

bronchial /'brɒŋkiəl/ *adj.* 支气管的

Brugia malayi /'brudʒiə mə'leɪaɪ/ *n.* 马来丝虫；马来布鲁线虫

caeca /'si:kə/ *n.* 盲肠（*sing.* caecum /'si:kəm/）

coelom /'si:ləm/ *n.* 体腔

corona /kə'rəʊnə/ *n.* 电晕；日冕；冠状物

cosmopolitan /ˌkɒzmə'pɒlɪtən/ *adj.* 世界性的

debilitating /dɪ'bɪlɪteɪtɪŋ/ *adj.* 使衰弱的

Digononta /daɪ'gəʊnəntə/ *n.* 双巢纲；双巢目

endoparasite /ˌendəʊ'pærəˌsaɪt/ *n.* 体内寄生虫

estuarine /'estjʊəˌraɪn/ *adj.* 河口的，江口的

filamentous /filə'mentəs/ *adj.* 丝状的

gastrotrich /'gæstrətrɪk/ *n.* 腹毛动物

genital /'dʒenɪtl/ *n.* 生殖器；外阴部；*adj.* 生殖的

gonochoristic /gɒ:nəʊ'kɔ:rɪstɪk/ *adj.* 雌雄异体的

hemocoel /'hi:məsi:l/ *n.* 血腔

hemolymph /'hi:məlɪmf/ *n.* 血淋巴

herbivore /'hɜ:bɪvɔ:r/ *n.* 食草动物，植食动物

hookworm /'hʊkwɜ:m/ *n.* 钩虫；钩虫病

introvert /'ɪntrəvɜ:t/ *n.* 翻吻动物

loiasis /ləʊ'aɪəsis/ *n.* 罗阿丝虫病；罗阿丝虫

lorica /lə'raɪkə/ *n.* 兜甲

lumen /'lu:mɪn/ *n.*（管状器官内的）内腔

mastax /'mæstæks/ *n.*（动物的）咀嚼囊

mesoderm /'mesə(ʊ)dɜ:m/ *n.* 中胚层；中胚叶

mollusc /'mɒləsk/ *n.* 软体动物

Monogononta /məˌnəʊgə'nɒntə/ *n.* 单巢目；单巢纲

moss /mɒs/ *n.* 苔藓；泥沼；*v.* 使长满苔藓

moult /məʊlt/ *v.* 脱毛，换羽；（毛、羽）蜕去

nematode /'nemətəʊd/ *n.* 线虫，线虫类

oligochaete /'ɒlɪgəki:t/ *n.* 寡毛纲（环节）动物

oviposition /ˌəʊvipə'zɪʃən/ *n.* 产卵

parthenogenesis /ˌpɑ:θənəʊ'dʒenɪsɪs/ *n.* 单性生殖，孤雌生殖

peritoneum /ˌperɪtə'niəm/ *n.* 腹膜

planktonic /ˌplæŋk'tɒnɪk/ *adj.* 浮游的

pinworm /'pɪnwɜ:m/ *n.* 蛲虫

portal /'pɔ:tl/ *n.* 大门，入口；*adj.* 门静脉的；肝门的

posterior /pɒ'stɪəriə(r)/ *n.* 后部；臀部

pseudocoelomate /ˌpsju:dəu'si:ləmeɪt/ *adj.* 有假腔的；*n.* 有假腔的动物

rotifer /'rəʊtɪfə/ *n.* 轮虫；轮形动物

rotting /'rɒtiŋ/ *adj.* 正在腐坏的

setae /'si:ti:/ *n.* 刚毛（*sing.* seta /'si:tə/）

speciose /'speʃiəs/ *adj.* 物种丰富的

stout /staʊt/ *adj.* 肥胖的；结实的

subterminal /sʌb'tɜ:mɪn(ə)l/ *adj.* 近末端的

translucent /trænz'lu:snt/ *adj.* 半透明的；透明的

transparent /træns'pærənt/ *adj.* 透明的；显然的

trichinosis /ˌtrɪkɪ'nəʊsɪs/ *n.* 旋毛虫病

trichuriasis /ˌtrɪkjʊ'raɪəsɪs/ *n.* 鞭虫病

vegetation /ˌvedʒə'teɪʃn/ *n.* 植被；植物，草木

vermiform /'vɜ:mɪfɔ:m/ *adj.* 蠕虫状的，蛆形的

whipworm /'wɪpˌwɜ:m/ *n.* 鞭虫

◆ Chapter 7

anchor /'æŋkə(r)/ *v.* 固定；抛锚；*n.* 锚

Annelida /ə'nelidə/ *n.* 环节动物门

annuli /'ænjʊˌlai/ 环形；轮（*sing.* annulus /'ænjʊləs/）

anticoagulant /ˌæntikəʊ'ægjələnt/ *n.* 抗凝血剂

brackish /'brækɪʃ/ *adj.* 含盐的；微咸的

brilliant /'brɪliənt/ *adj.* 灿烂的，闪耀的

capillary /kə'pɪləri/ *n.* 毛细血管；微血管

chaetae /'ki:ti:/ *n.* 刚毛；刺

chitinous /'kaɪtɪnəs/ *adj.* 壳质的；几丁质的

clitellum /klaɪ'teləm/ *n.* 环带；生殖带

compartment /kəm'pɑ:tmənt/ *v.* 分隔，划分；*n.* 隔室

conspicuous /kən'spɪkjuəs/ *adj.* 显著的；显而易见的

convolute /'kɒnvəˌlu:t/ *v.* 回旋；盘旋

copulation /ˌkɒpju'leɪʃn/ *n.* 交配；交尾；联结

courtship /'kɔ:tʃɪp/ *n.* 求爱；求婚；求爱期

crawl /krɔ:l/ *v.* 爬行；匍匐而行；*n.* 养鱼池

dimorphism /daɪ'mɔ:fɪzəm/ *n.* 二态性

earthworm /'ɜ:θwɜ:m/ *n.* 蚯蚓

ecdysozoan /ˌekdɪsəʊ'zəʊən/ *n.* 蜕皮动物

Errantia /'erəntiə/ *n.* 游走亚纲

esophagus /ɪ'sɒfəgəs/ *n.* 食管；食道

gizzard /'gɪzəd/ *n.* 砂囊；胃；喉咙

haemoglobin /ˌhi:mə'gləʊbɪn/ *n.* 血红蛋白；血色素

hermaphroditism /hɜ:'mæfrəˌdaɪtɪzəm/ *n.* 雌雄同体性

hirudin /'hɪrjʊdɪn/ *n.* 水蛭素

Hirudinea /ˌhiərju'diniə/ *n.* 蛭纲

humidity /hju:'mɪdəti/ *n.* 湿度；湿气

iridescent /ˌɪrɪ'desnt/ *adj.* 色彩斑斓的

jaw /dʒɔ:/ *n.* 口；狭口；咽喉

leech /litʃ/ *n.* 蛭类

Lumbricus /'ləmbrɪkəs/ *n.* 地龙，蚯蚓属；正蚓属

manoeuvre /mə'nuːvə(r)/ *n.* 演习；机动

metamerism /mɪ'tæmərɪzəm/ *n.* 位变异构；体节性；分节现象

metanephridia /ˌmetənə'frɪdɪə/ *n.* 后肾管

nephridia /nɪf'rɪdɪə/ *n.* 原肾（*sing.* nephridium /nɪf'rɪdɪəm/）

nephridiopore /ne'frɪdɪəpɔː/ *n.* 肾孔

paddle /'pædl/ *n.* 桨；*v.* 用桨划船

parapodia /ˌpærə'pəʊdɪə/ *n.* 疣足

photoreceptor /'fəʊtəʊrɪseptə(r)/ *n.* 光感受器

polychaete /'pɒlɪkiːt/ *n.* 多毛纲动物

proboscis /prə'bɒsɪs/ *n.* 鼻子；喙；象鼻；吻

prostomium /prəʊ'stəʊmɪəm/ *n.* 口前叶

protostome /'prəʊtəstəʊm/ *n.* 原肢类；原口动物

pygidium /paɪ'dʒɪdɪəm / *n.* 尾板；臀板

rhythmically /'rɪðmɪkli/ *adv.* 有节奏地

Sedentaria /ˌsedən'tæriə/ *n.* 隐居类；隐居亚纲

sedentary /'sedntri/ *adj.* 定栖的；定居的

septum /'septəm/ *n.* 隔膜；隔板

sheath /ʃiːθ/ *n.* 鞘；护套；叶鞘

sinuses /'saɪnɜːsɪs/ *n.* 鼻窦；鼻窦炎

spermatophore /'spɜːmətə(ʊ)fɔː/ *n.* 精囊；精原细胞

trochophore /'trəʊkə(ʊ)fɔː/ *n.* 担轮幼虫

undulation /ˌʌndju'leɪʃn/ *n.* 起伏；波动

valve /vælv/ *n.* 阀门；瓣膜；真空管

vas deferens /ˌvæs 'defərenz/ *n.* 输精管

victim /'vɪktɪm/ *n.* 受害者，牺牲者

◆ Chapter 8

abalone /ˌæbə'ləʊni/ *n.* 鲍鱼

ammonite /'æmənaɪts/ *n.* 菊石；鹦鹉螺

aorta /eɪ'ɔːtə/ *n.* 主动脉

auricle /'ɔːrɪkl/ *n.* 耳郭；外耳；心耳

belemnite /be'ləmnaɪt/ *n.* 箭石；乌贼的化石

bivalve /'baɪvælv/ *n.* 双壳贝；双壳类动物

buccal /'bʌk(ə)l/ *adj.* 颊的；口的，口腔的

calciferous /kæl'sɪfərəs/ *adj.* 含钙的；含碳酸钙的

camouflage /'kæməflɑ:ʒ/ *v.* 伪装，掩饰

chemoreceptor /'ki:məʊrɪseptə(r)/ *n.* 化学受体；化学感受器

chordate /'kɔ:deɪt/ *n.* 脊索动物；*adj.* 脊索动物的

chromatophore /'krəʊmətəfɔ:/ *n.* 色素体；载色体；色素细胞；叶绿体

cirri /'sɪraɪ/ *n.* 卷云；卷须；触毛（*sing.* cirrus /'sɪrəs/）

conch /kɒntʃ/ *n.* 贝壳；海螺壳

corpuscle /'kɔ:pʌsl/ *n.* 触觉小体

Crepidula /krepi'dju:lə/ *n.* 履螺属

ctenidium /tɪ'nɪdɪəm/ *n.* 栉鳃

cuttlefish /'kʌtlfɪʃ/ *n.* 乌贼；墨鱼

detorsion /dɪ'tɒrʃən/ *n.* 反扭转

escargot /eskɑr'gɒ/ *n.* 食用蜗牛

exhalant /eks'heɪlənt/ *adj.* 呼出的；蒸发的

gastropod /'gæstrəpɒd/ *n.* 腹足类动物

hemocyanin /ˌhi:mə'saɪənɪn/ *n.* 血蓝蛋白

hump /hʌmp/ *n.* 隆起；弓起

hydrostatic /ˌhaɪdrə(ʊ)'stætɪk/ *adj.* 静水力学的

intricate /'ɪntrɪkət/ *adj.* 复杂的；错综的

limpet /'lɪmpɪt/ *n.* 帽贝

mandatory /'mændətəri/ *adj.* 强制的；命令的

mantle /'mæntl/ *n.* 地幔；斗篷；覆盖物

Mesozoic /ˌmesəʊ'zəʊɪk/ *adj.* 中生代的；*n.* 中生代

mussel /'mʌsl/ *n.* 蚌；贻贝；淡菜

nautiloid /'nɔ:tɪlɔɪd/ *n.* 鹦鹉螺目软体动物

nautilus /'nɔ:tɪləs/ *n.* 鹦鹉螺

octopus /'ɒktəpəs/ *n.* 章鱼

opisthobranch /ə'pɪsθəbræŋtʃ/ *n.* 后鳃目软体动物

osphradium /ɒs'freidjəm/ *n.* 嗅检器，嗅觉器官

oyster /'ɔɪstə(r)/ *n.* 牡蛎，蚝

Paleozoic /ˌpæliə'zəuik/ *adj.* 古生代的；*n.* 古生代

palp /pælp/ *n.*（昆虫等的）触须

pericardium /ˌperɪ'kɑ:dɪəm/ *n.* 心包；心包膜

prosobranch /'prɒsəbræŋtʃ/ *n.* 前鳃亚纲软体动物

protandrous /prəʊ'tændrəs/ *adj.* 雄性先熟的

pulmonate /'pʌlmənеɪt/ *adj.* 有肺的；*n.* 有肺类动物

radula /'rædʒjʊlə/ *n.* 齿舌

relic /'relɪk/ *n.* 遗迹；遗骸；纪念物

slug /slʌg/ *n.* 蛞蝓

suprabranchial /'s(j)u:prə'bræŋkɪəl/ *adj.* 鳃上的

torsion /'tɔ:ʃn/ *n.* 扭转，扭曲；转矩

ultrafiltration /ˌʌltrəfɪl'treɪʃ(ə)n/ *n.* 超滤

ureteric /jʊ'ri:tə/ *adj.* 输尿管的

ventricle /'ventrɪkl/ *n.* 室；心室；脑室

visceral /'vɪsərəl/ *adj.* 内脏的

whelk /welk/ *n.* 海螺；峨螺

◆ Chapter 9

antennae /æn'teni:/ *n.* 触须；触角（*sing.* antenna /æn'tenə/）

aphid /'eɪfɪd/ *n.* 蚜虫

Arachnida /ə'ræknidə/ *n.* 蛛形纲

Araneae /ə'reini:/ *n.* 蜘蛛目

Arctic charr /'ɑ:ktik 'tʃɑ:(r)/ *n.* 北极红点鲑

bacteria /bæk'tɪəriə/ *n.* 细菌（*sing.* bacterium /bæk'tɪəriəm/）

barnacle /'bɑ:nəklz/ *n.* 藤壶

biramous /'bɪrəməs/ *adj.* 二枝的

Brachyura /ˌbræki'jurə/ *n.* 短尾亚目（甲壳类）

Branchiopoda /ˌbræŋki'ɒpədə/ *n.* 鳃足亚纲；鳃足类

Branchiura /'bræntʃiərə/ *n.* 鳃尾目；鳃尾纲；尾鳃蚓；鳃尾亚纲

carapace /'kærəpeɪs/ *n.* 壳；甲壳

caridina /kæri'dinə/ *n.* 米虾属；米虾

centipede /'sentɪpi:d/ *n.* 蜈蚣

Cephalocarida /ˌsefələu'kæridə/ *n.* 头虾纲；头虾亚纲；头甲亚目

cephalothorax /ˌsef(ə)ləʊ'θɔ:ræks/ *n.* 头胸；头胸部

Chelicerata /ˌkelə'serətə/ *n.* 有螯肢亚门；螯肢动物门

cheliped /'ki:ləped/ *n.* 螯足；带螯的前肢

Chilopoda /kai'lɒpədə/ *n.* 唇足亚纲；蜈蚣类

chitin /'kaɪtɪn/ *n.* 壳质；几丁质；角素；甲壳素

chrysalis /'krɪsəlɪs/ *n.* 蝶蛹；虫茧

cicada /sɪ'kɑ:də/ *n.* 蝉，知了

Cirripedia /'sɪrɪˌpediə/ *n.* 蔓足亚纲；蔓足纲

cockroach /'kɒkrəʊtʃ/ *n.* 蟑螂

Coleptera /ˌkoʊlɪ'ɑ:ptərə/ *n.* 鞘翅目；甲虫类

Copepoda /kəu'pepədə/ *n.* 桡足类

coxae /'kɒksi:/ *n.* 髋骨；髋关节；基节

crab /kræb/ *n.* 螃蟹

creepy /'kri:pi/ *adj.* 令人毛骨悚然的；爬行的

cricket /'krɪkɪt/ *n.* 板球（运动）；蟋蟀；蛐蛐

Crustacea /krʌs'teʃɪə/ *n.* 甲壳纲动物

cyclop /saɪk'lɒp / *n.* 剑水蚤

damselfly /'dæmzlflaɪ/ *n.* 豆娘（一种蜻蜓）；束翅亚目；蜻蜓目

Daphnia /'dæfniə/ *n.* 水蚤

Decapoda /də'kæpədə/ *n.* 十足目

denitrification /di:ˌnaɪtrɪfɪ'keɪʃn/ *n.* 脱氮；反硝化作用

Diaptomus /'diəptəməs/ *n.* 镖水蚤属

Diplopoda /di'pləpədə/ *n.* 倍足纲

Diptera /'dɪptərə/ *n.* 双翅目

dragonfly /'drægənflaɪ/ *n.* 蜻蜓

ecdysis /'ekdɪsɪs/ *n.*（动物的）蜕皮，换羽

ectoparasite /ektəʊ'pærəsaɪt/ *n.* 体表寄生虫

Eurypterida /juə'riptəridə/ *n.* 板足鲎亚纲

exocuticle /ek'səukju:tikəl/ *n.* 外角质层

femur /'fi:mə(r)/ *n.* 股骨；大腿骨

fishlice /'fɪʃlaɪs/ *n.* 鱼虱

flea /fli:/ *n.* 跳蚤

gnat /næt/ *n.* 蚊蚋；蚊子；小昆虫

gnathobase /'neɪθəʊbeɪs/ *n.* 鳃基；颚基

grasshopper /'grɑ:sˌhɒpə(r)/ *n.* 蝗虫；蚱蜢

grub /grʌb/ *n.*（昆虫的）幼虫；蛆；蛴螬

hemimetabolous /ˌhemɪmɪ'tæbələs/ *adj.* 半变态的

Hemiptera /hi'miptərə/ *n.* 半翅目

hind wing /haɪndwɪŋ/ *n.* 后翅

holometabolous /ˌhɒləmɪ'tæbələs/ *adj.*（完）全变态的

horseshoe /'hɔ:sʃu:/ *n.* 马蹄铁；U形物

Hymenoptera /ˌhaimi'nɒptərə/ *n.* 膜翅目

infraorder /ˌɪnfrə'ɔ:də/ *n.* 下目；次目

insect /'ɪnsekt/ *n.* 昆虫

labrum /'leɪbrəm/ *n.* 上唇

Lepidoptera /ˌlepɪ'dɑ:ptərə/ *n.* 鳞翅目

ligament /'lɪgəmənt/ *n.* 韧带；纽带，系带

Limulus polyphemus /'lɪmjʊləsˌpɔlə'fi:məs/ *n.* 美洲鲎

majestic /mə'dʒestɪk/ *adj.* 庄严的；宏伟的

Malacostraca /mælə'kɒstrəkə/ *n.* 软甲亚纲

Malpighian tubule /mæl'pigiən 'tju:bju:l/ *n.* 马氏管；马氏小管

mandible /'mændɪbl/ *n.* 下颌骨；上颚

meagre /'mi:gə(r)/ *adj.* 瘦的；贫弱的；贫乏的

meiobenthos /maiəu'benθɔs/ *n.* 较小型底栖生物

Merostomata /ˌmerəu'stəumətə/ *n.* 肢口纲

metameric /metə'merɪk/ *adj.* 位变异构的；分节的

metamorphosis /ˌmetə'mɔ:fəsɪs/ *n.* 变形；变质

Millipede /'mɪlɪpi:d/ *n.* 千足虫；倍足纲节肢动物

miniature /'mɪnətʃə(r)/ *adj.* 微型的，小规模的

monarch /'mɒnək/ *n.* 君主，帝王；最高统治者

mosquito /mə'ski:təʊ/ *n.* 蚊子

Mystacocarida /ˌmistə'kɔkəridə/ *n.* 须虾目；须鳃纲

nauplii /'nɔ:plɪaɪ/ *n.* 无节幼虫（*sing.* nauplius /'nɔ:plɪəs/）

nectar /'nektə(r)/ *n.* 花蜜；果汁饮料

nocturnal /nɒk'tɜ:nl/ *adj.*（动物的）夜间活动的

Odonata /ˌaʊdə'na:tə/ *n.* 蜻蜓目

oesophageal /i:sɒfə'dʒɪ:əl/ *adj.* 食道的

opisthosoma /əuˌpisθə'səumə/ *n.* 末体；后体；后体部；六角形的腹部

Orthoptera /ɔ:'θɔptərə/ *n.*（昆虫纲）直翅目

Ostracoda /'ɒstrəkɒdə/ *n.* 介形亚纲

outnumber /ˌaʊt'nʌmbə(r)/ *v.* 数目超过

Pauropoda /'pɔ:rəˌpɔdə/ *n.* 少脚纲；寡足纲

pedipalp /'pedɪpælp/ *n.* 须肢

pincer /'pɪnsə(r)/ *n.* 螯

pereon /pə'ri:ɒn/ *n.*（甲壳动物的）胸部

pheromone /'ferəməʊn/ *n.* 外激素；信息素

pollen /'pɒlən/ *n.* 花粉

pollinator /'pɑlɪneɪtə/ *n.* 传粉者，传粉媒介

pouch /paʊtʃ/ *n.*（有袋目动物腹部的）育儿袋

prawn /prɔ:n/ *n.* 对虾，明虾；*vi.* 捕虾

prolific /prə'lɪfɪk/ *adj.* 多产的；丰富的

pupa /'pju:pə/ *n.* 蛹

ramus /'reɪməs/ *n.* 分支；枝；羽支

receptacle /rɪ'septəkl/ *n.* 容器

rectum /'rektəm/ *n.* 直肠

retractile /rɪ'træktl/ *adj.* 可收缩的；回缩性

scorpion /'skɔ:piən/ *n.* 蝎子

seminiferous /ˌsemɪ'nɪf(ə)rəs/ *adj.* 输精的，结种子的

spermatid /'spɜ:mətɪd/ *n.* 精子细胞，精细胞

spermatozoa /ˌspə:mətə(ʊ)'zəʊə/ *n.* 精子

spiracle /'spaɪrək(ə)l/ *n.* 喷水孔；气门；呼吸孔

Symphyla /'simfilə/ *n.* 综合纲

tagmata /'tægmətə/ *n.* 体段，体节（*sing.* tagma /tægmə/）

tarsal /'tɑ:sl/ *adj.* 跗骨的；眼睑软骨的；*n.* 跗骨

thorax /'θɔ:ræks/ *n.* 胸，胸膛；胸腔

tibia /'tɪbiə/ *n.* 胫骨；胫节（昆虫）

trachea /trə'ki:ə/ *n.* 气管

tridimensional /ˌtraɪdɪ'menʃənəl/ *adj.* 立体的

Trilobite /'traɪləʊbaɪt/ *n.* 三叶虫

trochanter /trə'kæntə/ *n.* 转子；昆虫的转节

ultraviolet /ˌʌltrə'vaɪələt/ *adj.* 紫外的；紫外线的

venom /'venəm/ *n.* 毒液；*v.* 使有毒；放毒

waterproof /'wɔ:təpru:f/ *adj.* 防水的，不透水的

woodlice /'wʊdlais/ *n.* 木虱；潮虫（*sing.* woodlouse /'wɔdˌlaʊs/）

Xiphosura /'sifəsju:rə/ *n.* 剑尾类动物，剑尾目

◆ Chapter 10

aboral /æb'ɔ:r(ə)l/ *adj.* 对口的；离口的；*n.* 反口面

ambulacral /æmbjʊ'leɪkrəl/ *adj.* 步带的

ampullae /æm'pʊli:/ *n.* 壶腹

analogous /ə'næləgəs/ *adj.* 类似的；同功的；可比拟的

Asteroidea /'æstəˌrɔidiə/ *n.* 海星纲

Astropecten /ˌæstrəʊ'pekten/ *n.* 槭海星属

Astrophyton /ˌæstrəʊ'faiten / *n.* 筐蛇尾属

bifurcate /'baɪfəkeɪt/ *v.*（路、河等）分为两支；（使）分岔；*adj.* 分叉的；分支的

brittle /'brɪtl/ *adj.* 易碎的，脆弱的

bursae /'bə:si:/ *n.* 滑囊；黏液囊（*sing.* bursa /'bɜ:sə/）

Cidaris /'sidæris/ *n.* 头帕海胆属

cloacal /k'ləʊeɪkəl/ *adj.* 泄殖腔的

Crinoidea /krɪ'nɔidjə/ *n.* 海百合纲

dimorphic /daɪ'mɒrfɪk/ *adj.* 二态的

Echinocardium /i'kainəu'kɑ:rdiəm/ *n.* 心形海胆属

Echinodermata /i'kainə'də:mətə/ *n.* 棘皮动物门

Echinoidea /ˌekə'nɔidjə/ *n.* 海胆纲

Holothuroidea /ˌhɔləu'θurəidiə/ *n.* 海参纲；海参类

laden /'leɪdn/ *adj.* 负载的；装满的

madreporite /'mædrəpəˌraɪt/ *n.*（棘皮动物的）筛板；穿孔板

Mesothuria /'mesəuθiə/ *n.* 间海参属

Ophiothrix /ˌɒfiə'θriks/ *n.* 刺蛇尾属

Ophiuroidea /ˌɔfi'u:rɔidiə/ *n.* 蛇尾纲

ossicle /'ɒsɪk(ə)l/ *n.* 小骨，听小骨

papulae /'pæpjʊli:/ *n.* 丘疹（*sing.* papula /pæpju:lə/）

pedicellariae /ˌpedisə'læriə/ *n.* 叉棘

pentaradial /'pentəˌreɪdiəl/ *adj.* 五角形的

pigment /'pɪgmənt/ *n.* 色素；颜料；*vt.* 给……着色；*vi.* 呈现颜色

pinnule /'pɪnju:l/ *n.* 二回羽叶

podocyte /pɒdə'saɪt/ *n.* 足状突细胞

regenerate /rɪ'dʒenəreɪt/ *vt.* 使再生；革新

trauma /'trɔ:mə/ *n.* 精神创伤；外伤

urchin /'ɜ:tʃɪn/ *n.* 海胆

vascular /'væskjələ(r)/ *adj.* 血管的

vivid /'vɪvɪd/ *adj.* 生动的；鲜明的；鲜艳的

◆ Chapter 11

Amphibia /æm'fɪbɪə/ *n.* 两栖纲；两栖类

Ascidia /ə'sɪdɪə/ *n.* 海鞘属

atrium /'eɪtriəm/ *n.* 心房

Branchiostoma /ˌbræŋki'ɒstəmə/ *n.* 文昌鱼属；文昌鱼

cardiac /'kɑ:diæk/ *adj.* 心脏的，心脏病的；贲门的

cartilage /'kɑ:tɪlɪdʒ/ *n.* 软骨

cellulose /'seljuləʊs/ *n.* 纤维素；（植物的）细胞膜质

Cephalochordata /ˌsɛfələ(ʊ)kɔ:'deɪtə/ *n.* 头索动物亚门

cephalization /ˌsefəlaɪ'zeɪʃ(ə)n/ *n.* 头向集中，头部形成；头部集中化

cranium /'kreɪniəm/ *n.* 颅；头盖骨

Cyclostomata /ˌsaikləu'stəumətə/ *n.* 圆口亚纲；圆口类

Doliolum /'dɒliəʊləm/ *n.* 樽海鞘属

endocrine /'endəʊkrɪn/ *adj.* 内分泌（腺）的；*n.* 内分泌；内分泌腺；内分泌物；激素

hagfish /'hægfɪʃ/ *n.* 八目鳗；盲鳗

lancelet /'lɑ:nslɪt/ *n.* 文昌鱼

notochord /'nəʊtə(ʊ)kɔ:d/ *n.* 脊索

Pisces /'paɪsi:z/ *n.* 鱼纲

Salpa /'sælpə/ *n.* 樽海鞘属

slit /slɪt/ *n.* 狭长的切口；裂缝；撕裂

Tunicata /'tjʊnɪkɑ:tə/ *n.* 被囊亚门；被囊类

tunicin /'tjʊnɪsin/ *n.* 动物纤维素，皮素

unisexual /ju:nɪ'seksjʊəl/ *adj.* 单性的；雌雄异体的；雌雄异花的

Urochordata /ˌjuərəukɔ:'deitə/ *n.* 尾索动物亚门

Vertebrata /ˌvə:tə'breitə/ *n.* 脊椎动物亚门

◆ Chapter 12

Agnatha /'ægnəθə/ *n.* 无颌纲；无颚类脊椎动物

arch /ɑ:tʃ/ *n.* 弓形；拱门；拱起

branchial /'bræŋkɪəl/ *adj.* 鳃的；鳃状的

cranial /'kreɪniəl/ *adj.* 颅的，与颅骨有关的

cyclostome /'saɪklə(ʊ)stəʊm/ *n.* 圆口纲脊椎动物；*adj.* 圆口动物的

eel /iːl/ *n.* 鳗鱼；鳝鱼

horny /'hɔːni/ *adj.* 角的；角状的

jawless /'dʒɔːləs/ *adj.* 无颌的

lamprey /'læmpri/ *n.* 七鳃鳗；八目鳗

mesonephric /ˌmesəu'nefrik/ *adj.* 中肾的

Myxiniformes /'miksiniˌfɔːmiz/ *n.* 盲鳗目

nostril /'nɒstrəl/ *n.* 鼻孔

Paramyxine /'pærəmiksin/ *n.* 副盲鳗属

Petromyzontiformes /'petrəumaizɔntiˌfɔːmiz/ *n.* 七鳃鳗目

◆ Chapter 13

Actinopterygii /ˌæktinɒptə' rɪdʒiaɪ/ *n.* 辐鳍鱼纲；辐鳍鱼亚纲

Aphetohyoidea /ˌæfetəʊ'haɪɒɪdɪə/ *adj.* 游舌鱼总科

articulate /ɑː'tɪkjuleɪt/ *v.* 用关节连接；使相互连贯；发音；清楚地讲话

Chondrichthyes /kɔn'drikθiəs/ *n.* 软骨鱼纲

ctenoid /'tiːnɒɪd/ *adj.* 栉状的，有栉鳞的

cycloid /'saɪklɒɪd/ *n.* 摆线，圆滚线；*adj.* 圆形的；情绪起伏不定的

denticle /dentik(ə)l/ *n.* 小齿；齿状装饰

Dipnoi /'dipnɔi/ *n.* 肺鱼目

erythrocyte /ɪ'rɪθrəsaɪt/ *n.* 红细胞

ganoid /'gænɒɪd/ *adj.*硬鳞的；光鳞的；*n.* 硬鳞鱼

meroblastic /merəʊ'blæstɪk/ *adj.* 不完全卵裂的

myotome /'maɪətəʊm/ *n.* 肌节

operculum /ə(ʊ)'pɜːkjʊləm/ *n.* 鳃盖

Osteichthyes /ˌɔsti'ikθiiːz/ *n.* 硬骨鱼总纲

oviparous /əʊ'vɪpərəs/ *adj.* 卵生的；产卵的

ovoviviparous /ˌəʊvəʊvɪ'vɪpərəs/ *adj.* 卵胎生的

pancreas /'pæŋkriəs/ *n.* 胰腺

pectoral /'pektərəl/ *adj.* 胸的

pelvic /'pelvik/ *adj.* 骨盆的

Placodermi /'plækəˌdərmi/ *n.* 盾皮鱼纲

placoid /'plækɒɪd/ *adj.* 鱼鳞（板状的）；*n.* 有盾鳞的鱼

poikilothermous /ˌpɔikiləu'θəːməs/ *adj.* 变温的；冷血的；冷血动物的（=poikilothermal）

propulsion /prə'pʌlʃn/ *n.* 推进；推进力

Sarcopterygii /ˌsɑ:kɒptə'rɪdʒiaɪ/ *n.* 肉鳍鱼类
streamline /'stri:mlaɪn/ *adj.* 流线型的
tympanic /tɪm'pænɪk/ *adj.* 鼓膜的；鼓室的；鼓皮似的
ureotelic /ˌjʊərɪəʊ'telɪk/ *adj.* 排尿素的；排尿素代谢的
venous /'vi:nəs/ *adj.* 静脉的；有脉纹的

◆ Chapter 14

acrodont /'ækrədɒnt/ *n.* 端生齿
aestivate /'estɪveɪt/ *n.* 夏眠，夏蛰
Alytes /'ælits/ *n.* 产婆蟾属
Ambystoma /æm'bistəmə/ *n.* 钝口螈属；虎螈
ammonia /ə'məʊniə/ *n.* 氨
Amphiuma /æm'fɪəmə/ *n.* 三指螈属
arteriosus /'ɑ:təriəsəs/ *adj.* 动脉的
bladder /'blædə(r)/ *n.* 膀胱；囊状物
biconvex /baɪ'kɒnveks/ *adj.*（镜片等）两面凸的
buccopharyngeal /bʌkəfə'rɪndʒi:əl/ *adj.* 口咽的
Bufo /'bju:fəʊ/ *n.* 蟾蜍属
caecilian /sɪ'sɪlɪən/ *n.* 蚓螈
carboniferous /ˌkɑ:bə'nɪfərəs/ *adj.* 石炭纪的
copulatory /'kɑ:pjələˌtoʊri/ *adj.* 交配的
crocodile /'krɒkədaɪl/ *n.* 鳄鱼
dicondylic /daɪkɒn'dɪlɪk/ *adj.* 双突的
hepatic /hɪ'pætɪk/ *adj.* 肝的；肝脏色的
hoof /hu:f/ *n.*（马等动物的）蹄；人的脚
Hyla /'haɪlə/ *n.* 雨蛙属
Ichthyophis /ˌɪkθɪ'əʊsɪs/ *n.* 鱼螈属
limbless /lɪmlɪs/ *adj.* 无肢的；无足的
Necturus /'netrəs/ *n.* 泥螈属
newt /nju:t/ *n.* 蝾螈
occipital /ɒk'sɪpɪtəl/ *adj.* 枕骨的，枕部的；*n.* 枕骨
olfactory /ɒl'fæktəri/ *adj.* 嗅觉的；味道的；*n.* 嗅觉器官
pentadactyl /ˌpentə'dæktɪl/ *adj.* 五指的，五趾的
Pipa /pɪ'pɑ:/ *n.* 负子蟾属
pronephric /prəʊ'nefrɪk/ *adj.* 前肾的

protrusible /prəʊ'tru:səbl/ *adj*. 可伸出的；突出的

Rana /'rɑ:nə/ *n*. 蛙属

Rhacophorus /ˌrɪ'kɒfərəs/ *n*. 树蛙属

salamander /'sæləmændə(r)/ *n*. 火蜥蜴；蝾螈目动物

tadpole /'tædpəʊl/ *n*. 蝌蚪

Triturus /'tritju:rəs/ *n*. 蝾螈属，欧螈属

truncus /'trʌŋkəs/ *n*. 躯干

Tylototriton /'taɪləʊtəˌtraɪtən/ *n*. 疣螈属

Uraeotyphlus /ˌju:rɪəʊ'taɪfləs/ *n*. 盲尾蚓属

vocal /'vəʊkl/ *adj*. 声音的；有声的

vomerine /'voʊmərɪn/ *adj*. 犁骨的

Xenopus /zenə'pəs/ *n*. 爪蟾属

◆ Chapter 15

allantois /ə'læntɔɪs/ *n*. 尿囊；尿膜

alligator /'ælɪgeɪtə(r)/ *n*. 短吻鳄；美洲鳄；鳄鱼

ammonotelic /ˌæmɒ'nəʊtelɪk/ *adj*. 排氨的

amniote /'æmnioʊt/ *n*. 羊膜动物；脊椎动物

Anapsida /ə'næpsidə/ *n*. 缺弓亚纲；无孔亚纲；缺甲类

Archosauria /ɑ:kə'sɔ:riə/ *n*. 主龙类；初龙亚纲

Brontosaurus /ˌbrɒntə'sɔ:rəs/ *n*. 雷龙属

Bungarus /'bʌŋgərəs/ *n*. 环蛇属；金环蛇属

Calotes /kə'ləʊts/ *n*. 树蜥属

chameleon /kə'mi:liən/ *n*. 变色龙

Chelonia /kə'lonɪə/ *n*. 海龟属

chorion /'kɔ:rɪən/ *n*. 绒毛膜；浆膜；蛋壳

cobra /'kəʊbrə/ *n*. 眼镜蛇

Cotylosauria /ˌkɒtilə'sɔ:riə/ *n*. 杯龙目

Crocodilia /ˌkrɒkə'diliə/ *n*. 鳄目

Crocodylus /ˌkrɒkə'diləs/ 鳄属

desiccation /ˌdesɪ'keɪʃn/ *n*. 干燥

diaphragm /'daɪəfræm/ *n*. 隔膜；膈；隔板

Diapsida /daɪ'æpsɪdə/ *n*. 双孔亚纲

dinosaur /'daɪnəsɔ:(r)/ *n*. 恐龙

fossa /'fɒsə/ *n.* 小窝；凹

Gavialis /'gævaɪəlɪs/ *n.* 长吻鳄属

Hemidactylus /'hɛmiˌdæktləs/ *n.* 蜥虎属

Ichthyopterygia /ˌikθiˌɔptə'ridʒiə/ *n.* 鱼龙超目；鱼龙目

insectivorous /ˌɪnsek'tɪvərəs/ *adj.* 食虫的，以虫类为食的；食虫动植物的

krait /kraɪt/ *n.* 金环蛇

Lacertilia /ˌlæsə'tɪlɪə/ *n.* 蜥蜴亚目

lizard /'lɪzəd/ *n.* 蜥蜴；类蜥蜴爬行动物

macrolecithal /mækrəʊ'lesəθəl/ *adj.* 卵黄过多的；巨卵黄的

monocondylic /ˌmɒnəʊ'kɒndɪlic/ *adj.* 单接突的

Naja /'neidʒə/ *n.* 眼镜蛇属

Plesiosaurus /ˌplisɪə'sɔrəs/ *n.* 蛇颈龙属

polylecithal /'pɒliˌlesɪθl/ *adj.* 多卵黄的，多黄的

Reptile /'reptaɪl/ *n.* 爬行动物纲

Rhynchocephalia /'riŋkəuseˌfəlɪə / *n.* 喙头目

Sauria /'sɔ:riə/ *n.* 蜥蜴亚目；蜥蜴目

scute /skju:t/ *n.* 盾板；鳞甲

Seymouria /ˌsei'mɔ:rɪə/ *n.* 西蒙螈属

Sphenodon /'sfi:nədɒn/ *n.* 楔齿蜥属

Squamata /skwə'meitə/ *n.* 有鳞目

terrapin /'terəpɪn/ *n.* 水龟；泥龟；龟鳖

Testudo /tes'tjʊdo/ *n.* 陆龟属

tortoise /'tɔ:təs/ *n.* 龟；陆龟

Trionyx /'traɪɒnɪks/ *n.* 鳖属

tuatara /ˌtu:ə'tɑ:rə/ *n.* 大蜥蜴；喙头蜥

turtle /'tɜ:tl/ *n.* 龟；甲鱼；海龟

viper /'vaɪpə(r)/ *n.* 毒蛇；蝰蛇

Vipera /'vɪpərə/ *n.* 蝰蛇属

◆ Chapter 16

Archaeopteryx /ˌɑ:ki'ɒpterɪks/ *n.* 始祖鸟属；始祖鸟

Archaeornithes /ˌɑ:ki'ɔ:niθis/ *n.* 古鸟亚纲

forelimb /'fɔ:lɪm/ *n.* 前肢；前翼

gall /gɔ:l/ *n.* 胆汁；怨恨；苦味；*vt.* 烦恼；屈辱；磨伤；*vi.* 被磨伤

heron /'herən/ *n.* 鹭，苍鹭；深紫灰

hind /haɪnd/ *adj.* 后部的；*n.* 雌鹿

hop /hɒp/ *v.* 单足跳行；双足或齐足跳行；突然快速去某处；*n.* 单足短距离跳跃；双足短距离跳跃；短途旅行

kingfisher /'kɪŋfɪʃə(r)/ *n.* 翠鸟；鱼狗

Neornithes /ni'ɔ:niθiz/ *n.* 今鸟亚纲；新鸟亚纲；新鸟类；今鸟类

penguin /'peŋgwɪn/ *n.* 企鹅；空军地勤人员

perching /'pɜ:rtʃɪŋ/ *v.*（鸟）飞落；暂歇

spongy /'spʌndʒi/ *adj.* 海绵状的；轻软的；多孔而有弹性的；有吸水性的

syrinx /'sɪrɪŋks/ *n.* 鸣管；耳管

wade /weɪd/ *vi.* 跋涉；*vt.* 涉水；费力行走；*n.* 跋涉；可涉水而过的地方

◆ Chapter 17

antelope /'æntɪləʊp/ *n.* 羚羊

ape /eɪp/ *n.* 猿；*vt.* 模仿

armadillo /ˌɑrmə'dɪləʊ/ *n.* 犰狳

Artiodactyla /ˌɑ:tiəu'dæktilə/ *n.* 偶蹄类，偶蹄目

ass /æə/ *n.* 驴

blubber /'blʌbə(r)/ *n.* 鲸脂；哭泣；*vi.* 又哭又闹

camel /'kæml/ *n.* 骆驼；*adj.* 驼色的；暗棕色的

canine /'keɪnaɪn/ *adj.* 犬的；犬科的；似犬的；*n.* 犬；犬齿

cerebellum /serə'beləm/ *n.* 小脑

cerebrum /sə'ri:brəm/ *n.* 大脑

cervical /'sɜ:rvɪkl/ *adj.* 颈的；子宫颈的

Cetacea /sɪteɪʃ'jə/ *n.* 鲸下目；鲸类

cheetah /'tʃi:tə/ *n.* 猎豹

chisel /'tʃɪzl/ *n.* 凿子；*v.* 雕，刻，凿

cud /kʌd/ *n.* 反刍的食物

Dermoptera /də'mɒptərə/ *n.* 皮翼目

diastema /ˌdaɪə'sti:mə/ *n.* 间隙；间隙裂；纵裂

Edentata /ˌi:den'teitə/ *n.* 贫齿目

eutheria /'ju:θəriə/ *n.* 真兽亚纲

giraffe /dʒə'rɑ:f/ *n.* 长颈鹿

guinea pig /'gɪni pɪg/ *n.* 豚鼠；作为实验对象的人

hedgehog /'hedʒhɒg/ *n.* 刺猬

heterodont /'hetərədɒnt/ *adj.* 异形牙的；*n.* 异齿动物

hippopotamus /hɪpə'pɒtəməs/ *n.* 河马

hyaena /haɪ'i:nə/ *n.* 鬣狗

incisor /in'saɪər/ *n.* 门齿；切牙

Insectivora /ˌɪnsek'tɪvərə/ *n.* 食虫目；食虫类

insulator /'ɪnsjuleɪtə(r)/ *n.* 绝缘体；从事绝缘工作的工人

jackal /'dʒækl/ *n.* 豺

Lagomorpha /ˌlægə'mɒrfə/ *n.* 兔形目；兔类

lemur /'li:mər/ *n.* 狐猴

loris /'lɔ:rɪs/ *n.* 懒猴；蜂猴

Macropus /'mækrəʊpəs/ *n.* 大袋鼠属

mammary /'mæməri/ *adj.* 乳腺的；乳房的

marsupialia /mɑ:ˌsjupi'eiliə/ *n.* 有袋下纲

medulla /me'dʌlə/ *n.* 髓质

Metatheria /,metə'θiəriə/ *n.* 后兽次亚纲；后哺乳下纲

mongoose /'mɒŋgu:s/ *n.* 猫鼬；獴

monotremata /ˌmɒnəu'tri:mətə/ *n.* 单孔目

neocortex /ni:əʊ'kɔ:teks/ *n.* 新（大脑）皮质，皮层

opossum /ə'pɒsəm/ *n.* 负鼠

otter /'ɒtə(r)/ *n.* 水獭；水獭皮

pachyderm /'pækidɜ:m/ *n.* 厚皮类动物

pangolin /'pæŋgəlɪn/ *n.* 穿山甲；鲮鲤

parachute /'pærəˌʃu:t/ *n.* 降落伞；*vi.* 跳伞

patagium /pə'teɪdʒɪəm/ *n.* 翼膜

Perissodactyla /pəˌrɪso'dæktlə/ *n.* 奇蹄目；奇蹄类动物

Phascolarctos /ˌfɑ:skəʊ'lɑ:ktəʊs/ *n.* 树袋熊属；树袋熊

pinnae /'pɪni:/ *n.* 耳郭（*sing.* pinna /'pɪnə/）

placenta /plə'sentə/ *n.* 胎盘；胎座

porcupine /'pɒrkjəˌpaɪn/ *n.* 豪猪；箭猪

primate /praɪ'met/ *n.* 灵长类

proboscidea /ˌprəubə'sidiə/ *n.* 长鼻目；长鼻类

prosimian /pro'sɪmiən/ *adj.* 原猴亚目的；*n.* 原猴亚目的猴

Prototheria /ˌprəutəu'θi:riə/ *n.* 原哺乳亚纲；原兽亚纲

rhinoceros /raɪ'nɒsərəs/ *n.* 犀牛

rodentia /rəu'dentʃə/ *n.* 啮齿目；啮齿类

rumen /'ru:men/ *n.* 瘤胃（反刍动物的第一胃）

sea cow /'si:'kaʊ/ *n.* 海牛；海象

sebaceous /sɪ'beɪʃəs/ *adj.* 分泌脂质的；脂肪的，脂肪质的；似油脂或皮脂的

simian /'sɪmiən/ *adj.* 像猿（或猴）的；猿（或猴）的；*n.* 猿；猿猴；类人猿

Sirenia /saɪ'ri:niə/ *n.* 海牛目，海牛类

sloth /sləʊθ/ *n.* 怠惰，懒惰；树懒

sudoriferous /ˌs(j)u:də'rɪf(ə)rəs/ *adj.* 发汗的；分泌汗的

Tachyglossus /tækɪg'lɒsəs/ *n.* 针鼹属

tapir /'teɪpə(r)/ *n.* 貘

tarsier /'tɑ:sɪə/ *n.* 眼镜猴（多来源于东印度）；跗猴属动物

tusk /tʌsk/ *n.* 长牙；尖头，尖形物；*vt.* 以牙刺戳；以长牙掘

ungulate /'ʌŋgjələt/ *adj.* 有蹄的；像蹄子的；*n.* 有蹄类动物

vampire /'væmpaɪə(r)/ *n.* 吸血鬼；吸血蝙蝠

walrus /'wɔ:lrəs/ *n.* 海象

yak /jæk/ *n.* 牦牛；*vi.* 喋喋不休

◆ Chapter 18

Acanthodii /ˌækæn'θəudi:ai/ *n.* 棘鱼纲

anticlockwise /ˌænti'klɒkwaɪz/ *adj.* 逆时针的

armature /'ɑ:mətʃə(r)/ *n.* 盔甲，甲胄；防卫器官

armor /'ɑ:rmər/ *n.* 装甲防御

blastomere /'blæstə(ʊ)mɪə/ *n.* 分裂球；胚叶细胞

blastopore /'blæstəʊpɔ:/ *n.* 胚孔

brachiopod /'brækiəpɒd/ *n.* 腕足类动物

carpoid /kɑ:'pɔid/ *n.* 海果核类动物

Cenozoic /ˌsi:nə'zəuik/ *n.* 新生代；新生界的

chemosynthesis /ˌki:mə(ʊ)'sɪnθɪsɪs/ *n.* 化学合成

clade /kleɪd/ *n.* 分化枝；进化枝

clockwise /'klɒkwaɪz/ *adj.* 顺时针方向的

coelacanth /'si:ləkænθ/ *n.* 腔棘鱼

conodont /'kəʊnədɒnt/ *n.* 牙形石；牙形刺；牙形虫

crinoid /'kraɪnɒɪd/ *n.* 海百合纲动物

crossopterygian /krɒˌsɒptə'rɪdʒɪən/ *n.* 总鳍组鱼；总鳍鱼；总鳍鱼类

Ctenophora /ti'nɒfərə/ *n.* 栉水母动物门；栉水母类

delamination /diːˌlæmɪ'neɪʃən/ *n.* 分层；层离

deoxyribonucleic /diˌɑːksiˌraɪboʊnuː'kliːɪk/ *adj.* 脱氧核糖核酸的

descendant /dɪ'scndənt/ *n.* 后裔，子孙

deuterostome /'djuːtərəstoʊm/ *n.* 后口动物

Devonian /de'vəunjən/ *adj.* 泥盆纪的；*n.* 泥盆纪

dextral /'dekstr(ə)l/ *adj.* 右旋的；右侧的

diversification /daɪˌvɜːsɪfɪ'keɪʃn/ *n.* 多样化；变化

dominance /'dɒmɪnəns/ *n.* 优势；统治；支配

edentate /'iːdənteɪt/ *n.* 贫齿类动物；*adj.* 贫齿类的

enigmatic /ˌenɪg'mætɪk/ *adj.* 神秘的；高深莫测的

enormous /ɪ'nɔːməs/ *adj.* 庞大的，巨大的

Eocene /'iːəuiːn/ *adj.* 始新世的；*n.* 始新世

epiboly /ɪ'pɪbəlɪ/ *n.* 外包

extinct /ɪk'stɪŋkt/ *adj.* 灭绝的，绝种的

extirpate /'ekstəpeɪt/ *v.* 根除；彻底毁坏

fallopian /fə'loʊpiən/ *n.* 输卵管

flagellate /'flædʒəleɪt/ *n.* 鞭毛虫

forerunner /'fɔːrʌnə(r)/ *n.* 先驱；先驱者；预兆

gastraea /gæs'triːə/ *n.* 原肠祖

gastrulation /ˌgæstrʊ'leɪʃən/ *n.* 原肠胚形成

graptolite /'græptəlaɪt/ *n.* 笔石动物

ingression /ɪn'greʃən/ *n.* 进入；进入许可

innovation /ˌɪnə'veɪʃn/ *n.* 创新，革新；新方法

involution /ɪnvə'l(j)uːʃ(ə)n/ *n.* 卷入；内卷

Jurassic /dʒʊ'ræsɪk/ *adj.* 侏罗纪的；*n.* 侏罗纪

Latimeria /ˌlætɪ'mɪərɪə/ *n.* 矛尾鱼属；矛尾鱼

levotrophic /ˌlivəʊ'trəʊfɪk/ *adj.* 左旋的

marsupial /mɑː'suːpiəl/ *n.* 有袋类动物

Mississippian /ˌmɪsə'sɪpɪən/ *n.* 密西西比纪

mitotic /maɪ'tɒtɪk/ *adj.* 有丝分裂的；间接核分裂的

musculoskeletal /ˌmʌskjʊləʊ'skelɪt(ə)l/ *adj.* 肌（与）骨骼的

Neoproterozoic /niːoʊˌprotərə'zoɪk/ *n.* 新元古代

obliquely /ə'bliːkli/ *adv.* 倾斜地；转弯抹角地

oocyte /'əʊəsaɪt/ *n.* 卵母细胞

Ordovician /ˌɔːdəu'viʃən/ *n.* 奥陶纪；*adj.* 奥陶纪的

osseous /'ɒsiəs/ *adj*. 骨的；骨质的

Permian /'pə:miən/ *n*. 二叠纪；*adj*. 二叠纪的

phylogeny /faɪ'lɒdʒ(ə)nɪ/ *n*. 种系发生；系统发生

placental /plə'sentl/ *n*. 有胎盘哺乳动物；*adj*. 有胎盘的；胎盘的

pronuclei /prəʊ'nju:klɪˌaɪ/ *n*. 原核；前核（*sing*. pronucleus /prəʊ'nju:klɪəs/）

protochordate /ˌprəʊtəʊ'kɔ:deɪt/ *n*. 原索动物

pterobranch /ˌterə'bræŋk/ *n*. 羽鳃类

Pterosaur /'terəsɔ:/ *n*. 飞龙目；翼龙

rampant /'ræmpənt/ *adj*. 猖獗的；蔓延的；狂暴的；奔放的

recapitulate /ˌri:kə'pɪtʃuˌleɪt/ *n*. 概括；重述要点

Silurian /sɪ'ljʊərɪən/ *n*. 志留纪；*adj*. 志留纪的

Sinistral /'sɪnɪstr(ə)l/ *adj*. 左旋的；向左的

skeleton /'skelɪtn/ *n*. 骨架；骨骼

surpass /sə'pɑ:s/ *v*. 超越；胜过，优于

Synapsida /si'næpsidə/ *n*. 单弓目；下孔类；下孔亚纲

Triassic /traɪ'æsɪk/ *adj*. 三叠纪的；*n*. 三叠纪

zoologist /zu'ɒlədʒɪst/ *n*. 动物学家